The Finite Difference Method
in Partial Differential Equations

A.R. Mitchell and D.F. Griffiths
Department of Mathematics,
University of Dundee, Scotland

A Wiley-Interscience Publication

JOHN WILEY & SONS
Chichester · New York · Brisbane · Toronto

British Library Cataloguing in Publication Data

Mitchell, Andrew Ronald
    The finite difference method in partial differential
    equations.
    1. Differential equation, Partial
    2. Nets (Mathematics)
    1. Title II. Griffiths, D.F.
515'.353   QA374   79-40646

ISBN 0 471 27641 3

Typesetting by MHL Typesetting Ltd. Coventry
Printed by Page Brothers Limited, Norwich

*To Ann,
Cate and Sarah*

Contents

Preface

The study of numerical methods for the solution of partial differential equations has enjoyed an intense period of activity over the last thirty years from both the theoretical and practical points of view. Improvements in numerical techniques, together with the rapid advances in computer technology, have meant that many of the partial differential equations from engineering and scientific applications which were previously intractable can now be routinely solved.

This book is primarily concerned with finite difference techniques and these may be thought of as having evolved in two stages. During the fifties and early sixties many general algorithms were produced and analysed for the solution of standard partial differential equations. Since then the emphasis has shifted toward the construction of methods for particular problems having special features which defy solution by more general algorithms. This approach often necessitates a greater awareness of the different physical backgrounds of the problems such as free and moving boundary problems, shock waves, singular perturbations and many others particularly in the field of fluid dynamics. The present volume attempts to deal with both aspects of finite difference development with due regard to non-linear as well as linear differential equations. Often the solution of the sparse linear algebraic equations which arise from finite difference approximations forms a major part of the problem and so substantial coverage is given to both direct and iterative methods including an introduction to recent work on preconditioned conjugate gradient algorithms.

Although finite element methods now seem to dominate the scene, especially at a research level, it is perhaps fair to say that they have not yet made the impact on hyperbolic and other time-dependent problems that they have achieved with elliptic equations. We have found it appropriate to include an introduction to finite element methods but, in the limited space available, have concentrated on their relationships to finite difference methods.

The book is aimed at final year undergraduate and first year post-graduate students in mathematics and engineering. No specialized mathematical knowledge is required beyond what is normally taught in undergraduate courses in calculus and matrix theory. Although only a rudimentary knowledge of partial differential equations is assumed, anything beyond this would seriously limit the usefulness of the book, the dangers of developing numerical methods in ignorance of the corresponding theory cannot be emphasized too strongly. Theorems and proofs of existence, uniqueness, stability and convergence are seldom given and the reader is referred to appropriate research papers and advanced texts.

The sections devoted to applications reflect the strong links with computational fluid dynamics and it is hoped that practitioners in this field will find this material useful. It is taken for granted that the reader will have access to a digital computer since we believe that a proper understanding of the many methods described, along with their limitations, will be improved greatly by practical experience.

The list of references, which we readily admit cannot do justice to the vast literature in this field, is intended for the reader who wishes to pursue the subject at greater depth. We apologize to those many authors who have made important contributions in this area and whose names have not been mentioned. Most of the material in this book has been presented in the form of lectures to Honors and M.Sc. students in the Universities of Dundee and St Andrews. An earlier version of this text, written by one of the present authors, was published under the title *Computational Methods in Partial Differential Equations,* John Wiley and Sons, 1969.

In preparing the material for this book the authors have benefited greatly from discussions with many colleagues and former students. Special thanks are due to Ian Christie, Graeme Fairweather, Roger Fletcher, Sandy Gourlay, Pat Keast, Jack Lambert, John Morris and Bill Morton. Final thanks are due to Ros Hume for her swift and accurate typing of the manuscript.

A.R. MITCHELL
D.F. DRIFFITHS

Chapter 1

Basic Linear Algebra

1.1 Introduction

In the numerical solution of partial differential equations by finite difference methods, the differential system is replaced by a matrix system where the matrix A is usually *square* with *real* elements. In the present chapter some useful properties of the matrix A are outlined, often without proof. For more detailed information concerning properties of matrices, the reader is referred to books such as Fox (1964) Wilkinson (1965) and Fadeeva (1959).

The system of linear equations requiring solution is

$$\sum_{j=1}^{n} a_{ij}x_j = b_i \ (i = 1,2,\ldots,n), \tag{1}$$

which may be written as the matrix system

$$A\mathbf{x} = \mathbf{b}, \tag{2}$$

where A has n rows and columns and the elements $a_{ij}(i,j = 1,2,\ldots,n)$ are real numbers. The vectors $\mathbf{x}$ and $\mathbf{b}$ have n components.

The usual problem is to find $\mathbf{x}$ when A and $\mathbf{b}$ are given. A unique solution of equation (2) which may be written in the form

$$\mathbf{x} = A^{-1}\mathbf{b}$$

exists for equation (2), when A is non-singular, which is equivalent to A having a non-vanishing determinant. Since equation (2) is a matrix representation of a differential system, the matrix A is usually *sparse* (many of its elements are zero) and possesses a definite structure (determined by its non-zero elements). The method of inversion of A, particularly when the order n of the matrix is large, depends very much on the structure of A, and a variety of techniques for inverting A will be presented throughout this book. As n becomes larger, the methods of inversion must become more efficient.

1

2

A review of notation and properties for a square matrix A of order n with real elements, which is relevant to the solution of equation (1), is now given.

1.2 Notation

A — square matrix of order n.

a_{ij} — number in the real field, which is the element in the ith row and jth column of the matrix A.

A^{-1} — inverse of A.

A^T — transpose of A.

$|A|$ — determinant of A.

$\rho(A)$ — spectral radius of A.

I — unit matrix of order n.

0 — null matrix.

$\mathbf{x}$ — column vector with elements x_i $(i = 1,2,\ldots,n)$.

$\mathbf{x}^T$ — row vector with elements x_j $(j = 1,2,\ldots,n)$.

$\bar{\mathbf{x}}$ — complex conjugate of $\mathbf{x}$.

$\|A\|$ — norm of A.

$\|\mathbf{x}\|$ — norm of $\mathbf{x}$.

Π — permutation matrix which has entries of zeros and ones only, with one non-zero entry in each row and column.

1.3 Definitions

The matrix A is

non-singular if $|A| \neq 0$.

symmetric if $A = A^T$.

orthogonal if $A^{-1} = A^T$.

null if $a_{ij} = 0$ $(i,j = 1,2,\ldots,n)$.

diagonal if $a_{ij} = 0$ $(i \neq j)$.

diagonally dominant if $|a_{ii}| \geqslant \sum_{j \neq i} |a_{ij}|$ for all i.

tridiagonal if $a_{ij} = 0$ for $|i - j| > 1$.

block diagonal if

$$A = \begin{bmatrix} B_1 & & & & \\ & B_2 & & & \\ & & \cdot & & \\ & & & \cdot & \\ & & & & B_s \end{bmatrix}$$

where each $B_k (k = 1, 2, \ldots, s)$ is a square matrix, not necessarily of the same order.

upper triangular if $a_{ij} = 0, i > j$.
lower triangular if $a_{ij} = 0, j > i$.
irreducible if there exists no permutation transformation $\Pi A \Pi^{-1}$
   which reduces A to the form

$$\begin{bmatrix} P & 0 \\ R & Q \end{bmatrix}$$

where P and Q are square submatrices of order p and q respectively $(p + q = n)$
and 0 is a $p \times q$ null matrix.
The characteristic equation of A is

$$|A - \lambda I| = 0. \tag{3}$$

The eigenvalues of A are the roots $\lambda_i (i = 1, 2, \ldots, n)$ of the characteristic
   equation.
A right* eigenvector $\mathbf{x}^{(i)}$ for each λ_i is given by

$$A\mathbf{x}^{(i)} = \lambda_i \mathbf{x}^{(i)} \quad (\mathbf{x}^{(i)} \neq 0).$$

A left eigenvector $\mathbf{y}^{(i)T}$ for each λ_i is given by

$$\mathbf{y}^{(i)T} A = \lambda_i \mathbf{y}^{(i)T} \quad (\mathbf{y}^{(i)} \neq 0).$$

Two matrices A and B are similar if $B = H^{-1} AH$ for some non-singular matrix H.
$H^{-1} AH$ is a similarity transformation of A.
A and B commute if $AB = BA$.

Example 1   *Find the values of λ for which the set of equations*

$$\begin{aligned} (1 - \lambda)x_1 + x_2 - x_3 &= 0 \\ x_1 + (2 - \lambda)x_2 + x_3 &= 0 \\ x_1 + x_2 + (3 - \lambda) x_3 &= 0 \end{aligned}$$

has a non-zero solution and find one such solution.

A non-zero solution exists only if

$$\begin{vmatrix} 1 - \lambda & 1 & -1 \\ 1 & 2 - \lambda & 1 \\ 1 & 1 & 3 - \lambda \end{vmatrix} = 0,$$

which leads to
$$\lambda = 2, 2 \pm \sqrt{2}.$$

---

* The word 'right' is often omitted and $\mathbf{x}^{(i)}$ is usually referred to an an eigenvector corresponding to λ_i.

For $\lambda = 2$, the equations become

$$
\begin{array}{lrl}
\text{(a)} & -x_1 + x_2 - x_3 & = 0, \\
\text{(b)} & x_1 + x_3 & = 0, \\
\text{(c)} & x_1 + x_2 + x_3 & = 0.
\end{array} \Bigg\} \tag{4}
$$

Only two of the equations of (4) are independent. For example, (b) can be obtained by halving the difference of (c) and (a). If we ignore (a) it follows from (b) and (c) that

$$ x_2 = 0 \text{ and } x_1/x_3 = -1. $$

Thus any eigenvector corresponding to $\lambda = 2$ is proportional to

$$
\begin{bmatrix} 1 \\ 0 \\ -1 \end{bmatrix}.
$$

Example 2 If $\mathbf{x}^{(i)}$ ($i = 1, 2, \ldots, n$) are the eigenvectors of A and $\mathbf{y}^{(j)}$ ($j = 1, 2, \ldots, n$) are the eigenvectors of A^T, show that

$$ \mathbf{x}^{(i)T}\mathbf{y}^{(j)} = 0 \quad (\lambda_i \neq \lambda_j), $$

where λ_i ($i = 1, 2, \ldots, n$) are the eigenvalues of A.

The eigenvalues of A^T are given by

$$ |A^T - \lambda I| = 0, $$

since the eigenvalues of A^T are the same as those of A. An eigenvector of A^T corresponding to λ_j is $\mathbf{y}^{(j)}$, given by

$$ A^T\mathbf{y}^{(j)} = \lambda_j\mathbf{y}^{(j)}, \tag{5} $$

or after transposing both sides,

$$ \mathbf{y}^{(j)T}A = \lambda_j\mathbf{y}^{(j)T}. $$

($\mathbf{y}^{(j)T}$ is a left eigenvector of A.)
Also

$$ A\mathbf{x}^{(i)} = \lambda_i\mathbf{x}^{(i)}, $$

which, on transposing both sides, gives

$$ \mathbf{x}^{(i)T}A^T = \lambda_i\mathbf{x}^{(i)T}. \tag{6} $$

Postmultiplying equation (6) by $\mathbf{y}^{(j)}$ and premultiplying equation (5) by $\mathbf{x}^{(i)T}$, and subtracting, the result

$$ 0 = (\lambda_i - \lambda_j)\mathbf{x}^{(i)T}\mathbf{y}^{(j)} $$

is obtained. This leads to the desired result if $\lambda_i \neq \lambda_j$.

Example 3 Show that the eigenvalues of a matrix are preserved under a similarity transformation.

If

$$Ax = \lambda x, \qquad x \neq 0$$

then

$$H^{-1} Ax = \lambda H^{-1}x \quad (|H| \neq 0),$$

leading to

$$H^{-1} AHH^{-1}x = \lambda H^{-1}x,$$

and so

$$(H^{-1} AH)H^{-1}x = \lambda H^{-1}x.$$

Thus the eigenvalues are preserved and the eigenvectors are multiplied by H^{-1}.

Example 4 If the eigenvalues of the matrix A are distinct, show that there is a similarity transformation which reduces A to diagonal form and has its columns equal to the eigenvectors of A.

A has eigenvalues $\lambda_1, \lambda_2, \ldots, \lambda_n$ and eigenvectors $x^{(1)}, x^{(2)}, \ldots, x^{(n)}$, A^T has eigenvalues $\lambda_1, \lambda_2, \ldots, \lambda_n$ and eigenvectors $y^{(1)}, y^{(2)}, \ldots, y^{(n)}$, where

$$y^{(j)T}x^{(i)} = 0 \quad (i \neq j),$$

and

$$y^{(i)T}x^{(i)} = 1 \quad (i = 1, 2, \ldots, n),$$

the eigenvectors having been normalized. These relations imply that the matrix Y^T which has $y^{(j)T}$ as its jth row is the inverse of the matrix X, which has $x^{(i)}$ as its ith column. Now

$$AX = X \operatorname{diag}(\lambda_i),$$

and since Y^T is the inverse of X, it follows that

$$X^{-1} AX = Y^T AX = \operatorname{diag}(\lambda_i),$$

leading to the desired result.

1.4 Linear vector space

An important role will be played by the n-dimensional vector space R_n. A point x in this space is arrayed in the form

$$\mathbf{x} = \begin{bmatrix} x_1 \\ x_2 \\ \cdot \\ \cdot \\ \cdot \\ x_n \end{bmatrix},$$

where $x_1, x_2, \ldots, x_n$ are the n components of the vector, which may be complex, but which we shall assume to be real. The number n is said to be the *dimension* of the space.

Vectors $\mathbf{x}^{(1)}, \mathbf{x}^{(2)}, \ldots, \mathbf{x}^{(n)}$ are said to be *linearly dependent* if non-zero constants $c_1, c_2, \ldots, c_n$ exist such that

$$c_1 \mathbf{x}^{(1)} + c_2 \mathbf{x}^{(2)} + \ldots + c_n \mathbf{x}^{(n)} = 0.$$

If this equation holds only for $c_1 = c_2 = \ldots = c_n = 0$, however, the vectors $\mathbf{x}^{(1)}$, $\mathbf{x}^{(2)}, \ldots, \mathbf{x}^{(n)}$ are said to be *linearly independent*. A system of linearly independent vectors is said to constitute a *basis* for a space, if any vector of the space is a linear combination of the vectors of the system. The number of vectors forming a basis is equivalent to the dimension of the space. The n linearly independent vectors form a complete system and are said to span the whole n space.

The *inner (or scalar) product* of two members $\mathbf{x}$ and $\mathbf{y}$ of the vector space is defined by

$$(\mathbf{x}, \mathbf{y}) = \sum_{i=1}^{n} x_i y_i.$$

The *length* of a vector $\mathbf{x}$ is given by

$$(\mathbf{x}, \mathbf{x})^{1/2} = \left(\sum_{i=1}^{n} x_i^2 \right)^{1/2}$$

The non-zero vectors $\mathbf{x}$ and $\mathbf{y}$ are said to be *orthogonal* if $(\mathbf{x}, \mathbf{y}) = 0$. A system of vectors is orthogonal if any two vectors of the system are orthogonal to one another.

Theorem The vectors forming an orthogonal system are linearly independent.

Proof: Let $\mathbf{x}^{(1)}, \mathbf{x}^{(2)}, \ldots, \mathbf{x}^{(n)}$ form an orthogonal system and suppose that

$$c_1 \mathbf{x}^{(1)} + c_2 \mathbf{x}^{(2)} + \ldots + c_n \mathbf{x}^{(n)} = 0.$$

Take the scalar product with $\mathbf{x}^{(i)}$, and so

$$c_i(\mathbf{x}^{(i)}, \mathbf{x}^{(i)}) = 0$$

for any $i = 1, 2, \ldots, n$. Since $(\mathbf{x}^{(i)}, \mathbf{x}^{(i)}) \neq 0$, it follows that

$$c_i = 0 \quad (i = 1, 2, \ldots, n).$$

Thus the vectors $\mathbf{x}^{(1)}, \mathbf{x}^{(2)}, \ldots, \mathbf{x}^{(n)}$ are linearly independent.

1.5 Useful matrix properties

As far as possible these properties are grouped, although there is no particular merit in the order of presentation chosen. All matrices are square of order n. We shall be concerned almost entirely with real matrices, but some of the results in this section apply when A is complex. This is particularly true of the Jordan canonical form where one usually requires to work with complex numbers.

Jordan canonical form

A Jordan submatrix of A is a matrix of the form

$$\begin{bmatrix} \lambda_i & & & & \\ 1 & \lambda_i & & \text{\huge O} & \\ & 1 & \cdot & & \\ & & \cdot & \cdot & \\ & & & \cdot & \cdot \\ & \text{\huge O} & & & \cdot & \cdot \\ & & & & 1 & \lambda_i \end{bmatrix}$$

where λ_i is an eigenvalue of A. The Jordan canonical form of A is a block diagonal matrix composed of Jordan submatrices. It is unique up to permutations of the blocks. Any matrix A can be reduced to Jordan canonical form by a similarity transformation

$$J = H^{-1}AH.$$

The diagonal elements of J are the eigenvalues of A.

If A has n distinct eigenvalues, its Jordan canonical form is diagonal and its n associated eigenvectors are linearly independent. They form a complete system of eigenvectors and span the whole n-dimensional space. If A does not have n distinct eigenvalues, it may or may not possess n independent eigenvectors.

If any two matrices A and B commute and have diagonal canonical forms, then they have a complete set of simultaneous eigenvectors.

Symmetric matrix

A symmetric matrix has

(i)   a diagonal Jordan canonical form;
(ii) n real eigenvalues; and
(iii) n mutually orthogonal eigenvectors.

If A and B are symmetric, and $AB = BA$, then AB is symmetric.

Positive definite matrix

If A is real and $\mathbf{x}$ is complex, then A is *positive definite* if

$$(\mathbf{x}, A\mathbf{x}) > 0 \text{ for all } \mathbf{x} \neq 0.$$

This, of course, implies that $(\mathbf{x}, A\mathbf{x})$ is real. If A is positive definite, then it is symmetric. (Note that the inner product $(\mathbf{x}, \mathbf{y})$ of two complex vectors is $\sum_{i=1}^{n'} x_i \bar{y}_i$, where $\bar{y}_i$ is the complex conjugate of y_i)

If A is real and $\mathbf{x}$ is real, then A is *positive real* if

$$(\mathbf{x}, A\mathbf{x}) > 0 \text{ for all } \mathbf{x} \neq 0.$$

This time A is not necessarily symmetric.

A matrix A is *positive semi-definite* if

$$(\mathbf{x}, A\mathbf{x}) \geq 0 \text{ for all } \mathbf{x},$$

with equality for at least one $\mathbf{x} \neq 0$.

A *Stieltjes matrix* is a real positive definite matrix with all its off-diagonal elements non-positive. If the properties of irreducibility and diagonal dominance are added, the matrix is often referred to as an S-matrix. An S-matrix has the following properties:

(i) $a_{ij} = a_{ji}$;
(ii) $a_{ij} \leq 0$ for $i \neq j$;
(iii) $|a_{ii}| \geq \sum_{j \neq i} |a_{ij}|$, with strict inequality for at least one i;
(iv) $S \equiv [a_{ij}]$ is irreducible;
(v) S is positive definite; and
(vi) the elements of S^{-1} are positive.

Such matrices occur repeatedly in the finite difference solution of partial differential equations.

Example 5 Show that, if $(\mathbf{x}, A\mathbf{x}) > 0$ for all complex $\mathbf{x}$, then A is symmetric.

Let $\mathbf{x} = \mathbf{a} + i\mathbf{b}$ where $\mathbf{a}$ and $\mathbf{b}$ are real. Now the inner product $(\mathbf{x}, \mathbf{y})$ of two complex vectors is

$$\sum_{i=1}^{n} x_i \bar{y}_i,$$

where $\bar{y}_i$ is the complex conjugate of y_i. Consequently,

$$(\mathbf{x}, A\mathbf{x}) = (\mathbf{a} + i\mathbf{b}, A(\mathbf{a} + i\mathbf{b}))$$
$$= (\mathbf{a}, A\mathbf{a}) + i(\mathbf{b}, A\mathbf{a}) - i(\mathbf{a}, A\mathbf{b}) + (\mathbf{b}, A\mathbf{b})$$
$$= [(\mathbf{a}, A\mathbf{a}) + (\mathbf{b}, A\mathbf{b})] - i[(\mathbf{a}, A\mathbf{b}) - (\mathbf{b}, A\mathbf{a})] > 0.$$

This is only possible if

$$(\mathbf{a}, A\mathbf{b}) - (\mathbf{b}, A\mathbf{a}) = (\mathbf{a}, A\mathbf{b}) - (\mathbf{a}, A^T\mathbf{b}) = (\mathbf{a}, (A - A^T)\,\mathbf{b}) = 0,$$

and so

$$A = A^T,$$

leading to the desired result.

Example 6 Show that if A is symmetric and positive real, then its eigenvalues are all positive.

If A is symmetric, its real eigenvalues imply real eigenvectors. Consequently,

$$(\mathbf{x}, A\mathbf{x}) = (\mathbf{x}, \lambda\mathbf{x}) = \lambda(\mathbf{x}, \mathbf{x})$$

for any eigenvector $\mathbf{x} \neq 0$. But

$$(\mathbf{x}, A\mathbf{x}) > 0,$$

since A is positive real, and so

$$\lambda = \frac{(\mathbf{x}, A\mathbf{x})}{(\mathbf{x}, \mathbf{x})} > 0.$$

Example 7 Show that $A^T A$ has non-negative eigenvalues.

Let

$$B = A^T A.$$

B is symmetric because

$$B^T = (A^T A)^T = A^T A = B,$$

and so B has real eigenvalues and real eigenvectors. For any real non-zero $\mathbf{x}$,

$$(\mathbf{x}, B\mathbf{x}) = (\mathbf{x}, A^T A\mathbf{x}) = (A\mathbf{x}, A\mathbf{x}) \geq 0,$$

and so B is positive semi-real and, by previous example, has non-negative eigenvalues.

Eigenvalues of a matrix

The eigenvalues of A lie within the union of the n discs

$$|z - a_{ii}| \leq \sum_{\substack{j=1 \\ j \neq i}}^{n} |a_{ij}| \quad (i = 1, 2, \ldots, n)$$

in the complex z plane. Since A^T has the same eigenvalues as A, this may be replaced by

$$|z - a_{jj}| \leqslant \sum_{\substack{i=1 \\ i \neq j}}^{n} |a_{ij}| \qquad (j = 1, 2, \ldots, n).$$

This is *Gerschgorin's theorem*.

The *spectral radius* of a matrix A is denoted by $\rho(A)$ and is given by

$$\rho(A) = \max_{i} |\lambda_i|,$$

where $\lambda_i (i = 1, 2, \ldots, n)$ are the eigenvalues of A. $\rho(A)$ is the radius of the smallest circular disc in the complex plane, with the centre as the origin, which contains all the eigenvalues of A. From Gerschgorin's theorem,

$$\rho(A) \leqslant \min \left(\max_{i} \sum_{j} |a_{ij}|; \max_{j} \sum_{i} |a_{ij}| \right).$$

If A is the tridiagonal matrix

$$\begin{bmatrix} a & b & & & & \\ c & a & b & & \text{\Large O} & \\ & \cdot & \cdot & \cdot & & \\ & & \cdot & \cdot & \cdot & \\ & & & \cdot & \cdot & b \\ & \text{\Large O} & & & c & a \end{bmatrix}$$

where a, b and c are real and $bc > 0$, the eigenvalues of A are given in closed form by

$$\lambda_s = a + 2\sqrt{(bc)} \cos \frac{s\pi}{n+1} \qquad (s = 1, 2, \ldots, n).$$

Exercise

1. Find the eigenvectors of the matrix

$$\begin{bmatrix} -2 & 1 & 0 \\ 1 & -2 & 1 \\ 0 & 1 & -2 \end{bmatrix},$$

and show that they are mutually orthogonal.

1.6 Vector and matrix norms

The modulus of a complex number gives an assessment of its overall size. It will be useful to have a single number which plays a similar role in the case of a vector or a matrix. Such a quantity will be called a *norm*.

The norm of a vector will be denoted by $\|x\|$ and it satisfies the following requirements:

(i) $\|x\| > 0$ for $x \neq 0$;

(ii) $\|kx\| = |k| \, \|x\|$ for any complex number k; and

(iii) $\|x + y\| \leqslant \|x\| + \|y\|$ (triangle inequality).

Example 8 Show that

$$\|x - y\| \geqslant |\, \|x\| - \|y\| \,|.$$

We have

$$\|x\| = \|x - y + y\| \leqslant \|x - y\| + \|y\|,$$

and so

$$\|x - y\| \geqslant \|x\| - \|y\|.$$

But

$$\|x - y\| = \|y - x\| \geqslant \|y\| - \|x\|,$$

leading to

$$\|x - y\| \geqslant |\, \|x\| - \|y\| \,|.$$

There are three simple ways of assigning a vector norm. They follow from

$$\|x\|_p = (|x_1|^p + |x_2|^p + \ldots + |x_n|^p)^{1/p} \quad (p = 1, 2, \infty),$$

where

$$x = \begin{bmatrix} x_1 \\ x_2 \\ \cdot \\ \cdot \\ x_i \\ \cdot \\ \cdot \\ x_n \end{bmatrix},$$

and $\|x\|_\infty$ is interpreted as $\max\limits_i |x_i|$ and is referred to as the *maximum* norm. It should be noted that the norm arising from $p = 2$ is none other than the length of the vector defined previously. It is usually referred to as the L_2 norm.

Exercise

2. Show that

   (i) $\|\mathbf{x}\|_2 \leqslant \sqrt{n}\|\mathbf{x}\|_\infty$,

   (ii) $\|\mathbf{x}\|_1 \leqslant n\|\mathbf{x}\|_\infty$.

In a similar manner, the norm of a square matrix A is a non-negative number denoted by $\|A\|$ satisfying the following conditions:

   (i) $\|A\| > 0$ if $A \neq 0$;

   (ii) $\|kA\| = |k|\,\|A\|$ for any complex number k;

   (iii) $\|A + B\| \leqslant \|A\| + \|B\|$; and

   (iv) $\|AB\| \leqslant \|A\|\,\|B\|$.

Since matrices and vectors appear simultaneously, it is convenient to introduce the norm of a matrix in such a way that it is compatible with a given vector norm. A matrix norm is said to be *compatible* with a given vector norm if

$$\|A\mathbf{x}\| \leqslant \|A\|\,\|\mathbf{x}\|, \tag{7}$$

for all non-zero $\mathbf{x}$. The formula which is used to construct the matrix norm *subordinate* to a given vector norm is

$$\|A\| = \max_{\|x\|=1} \|A\mathbf{x}\|, \tag{8}$$

which states that the norm of the matrix A is the maximum of the norms of the vectors $A\mathbf{x}$, where $\mathbf{x}$ is allowed to run over the set of all vectors whose norm is unity. This maximum is attainable for each matrix A, and so a vector $\mathbf{x}_0$ can be found such that

$$\|\mathbf{x}_0\| = 1$$

and

$$\|A\mathbf{x}_0\| = \|A\|.$$

If $A = I$, it follows from equation (7) that $\|I\| = 1$, which is the condition for any matrix norm to be subordinate to a given vector norm.

The matrix norm subordinate to $\|\mathbf{x}\|_p$ is denoted by $\|A\|_p$, and these norms satisfy the relations

$$\|A\|_1 = \max_j \sum_i |a_{ij}|,$$

$$\|A\|_2 = \max \{|\lambda|; \lambda \text{ an eigenvalue of } (A^T A)^{1/2}\},$$

and

$$\|A\|_\infty = \max_i \ \Sigma_j \ |a_{ij}|,$$

respectively. It follows that

$$\|A\|_1 = \|A^T\|_\infty$$

The proof of the result for the norm $\|A\|_2$, which is often called the *spectral norm*, is instructive and will be given later.

Exercise

3. (Varga [1962]). If

$$A = \begin{bmatrix} \alpha & 4 \\ 0 & \alpha \end{bmatrix},$$

where α is an arbitrary real parameter, show that

$$\|A^m\|_2 = \alpha^m \left[1 + 8 \frac{m^2}{\alpha^2} \left\{ 1 + \left(1 + \frac{\alpha^2}{4m^2} \right)^{1/2} \right\} \right]^{1/2}$$

1.7 Theorems relating matrix norm to spectral radius

Theorem I   For any matrix A, $\|A\| \geqslant \rho(A)$ where $\|A\|$ is any norm of A.

Proof: For any eigenvalue λ, and associated non-zero eigenvector $\mathbf{x}$ of A, it follows that

$$|\lambda| \ \|\mathbf{x}\| = \|\lambda\mathbf{x}\| = \|A\mathbf{x}\| \leqslant \|A\| \ \|\mathbf{x}\|,$$

and so

$$|\lambda| \leqslant \|A\| \quad \text{for any eigenvalue } \lambda \text{ of } A.$$

Thus

$$\rho(A) \leqslant \|A\|.$$

Theorem II   For any real matrix A, $\|A\|_2 = [\rho(A^T A)]^{1/2}$, where $\|A\|_2$ is the spectral norm of A.

Proof: $A^T A$ is symmetric and non-negative definite. Let $\{\mathbf{x}^{(i)}\}$ $(i = 1,2,\ldots,n)$ be an orthonormal set of real eigenvectors of $A^T A$, i.e.

with

$$A^T A x^{(i)} = \lambda_i x^{(i)} \ (0 \leqslant \lambda_1 \leqslant \lambda_2 \leqslant \ldots \leqslant \lambda_n),$$

$$x^{(i)T} x^{(j)} = 0 \ (i \neq j)$$

and

$$x^{(i)T} x^{(i)} = 1 \ (1 \leqslant i \leqslant n).$$

Any other non-zero vector x in the space spanned by $x^{(i)} \ (i = 1, 2, \ldots, n)$ can be expressed as

$$x = \sum_{i=1}^{n} c_i x^{(i)},$$

and so

$$\left(\frac{\|Ax\|_2}{\|x\|_2} \right)^2 = \frac{(Ax, Ax)}{(x, x)} = \frac{x^T A^T A x}{x^T x}$$

$$= \frac{(\sum_i c_i x^{(i)})^T \ (\sum_i \lambda_i c_i x^{(i)})}{(\sum_i c_i x^{(i)})^T \ (\sum_i c_i x^{(i)})} = \frac{\sum_i \lambda_i |c_i|^2}{\sum_j |c_j|^2}.$$

This gives the result

$$0 \leqslant \lambda_1 \leqslant \left(\frac{\|Ax\|_2}{\|x\|_2} \right)^2 \leqslant \lambda_n.$$

But $x = x^{(n)}$ shows that equality is possible on the right, and so

$$\|A\|_2^2 = \max_{\|x\|_2 \neq 0} \frac{\|Ax\|_2}{\|x\|_2} = \lambda_n = \rho(A^T A).$$

Theorem III If A is symmetric, $\|A\|_2 = \rho(A)$.

Proof : $\|A\|_2^2 = \rho(A^T A) = \rho(A^2) = \rho^2(A)$, and hence the result follows.

Exercises

4. Show that $\rho(A^2) \equiv \rho^2(A)$.
5. Find the spectral radius and spectral norm of

$$\begin{bmatrix} 1 & 0 & \alpha \\ 0 & 1 & 0 \\ 0 & 0 & 1 \end{bmatrix},$$

where α is a real parameter, and show that these are equal only if $\alpha = 0$.

6. If
$$A = \begin{bmatrix} 2 & 1 \\ 1 & 2 \end{bmatrix},$$

show that $\|A\|_2 = \rho(A)$, but if

$$A = \begin{bmatrix} 2 & 1 \\ 0 & 2 \end{bmatrix},$$

show that $\|A\|_2 > \rho(A)$.

7. Show that $\|A\|_2^2 \leqslant \|A\|_\infty \|A\|_1$.

1.8 Convergence of sequences of matrices

The matrix A is *convergent* to zero if the sequence of matrices $A, A^2, A^3, \ldots$ converges to the null matrix 0.

Theorem I $\lim_{r \to \infty} A^r = 0$ *if* $\|A\| < 1$.

Proof:

$$\|A^r\| = \|AA^{r-1}\| \leqslant \|A\| \|A^{r-1}\| \leqslant \|A\|^2 \|A^{r-2}\| \leqslant \bar{\ } \ldots \leqslant \|A\|^r,$$

and so the result follows.

Theorem II $\lim_{r \to \infty} A^r = 0$ *if and only if* $|\lambda_i| < 1$ *for all eigenvalues* λ_i ($i = 1, 2, \ldots, n$) *of* A.

Proof: Consider the Jordan canonical form of A. A Jordan submatrix of A is of the form

$$\begin{bmatrix} \lambda_i & & & & \\ 1 & \lambda_i & & \text{O} & \\ & 1 & \cdot & & \\ & & \cdot & \cdot & \\ & & & \cdot & \cdot \\ \text{O} & & & \cdot & \cdot \\ & & & & 1 & \lambda_i \end{bmatrix}$$

where λ_i is an eigenvalue of A. If the Jordan submatrix is raised to the power r, then the result tends to the null matrix as $r \to \infty$, if and only if $|\lambda_i| < 1$.

Example 9 If

$$A = \begin{bmatrix} 0.8 & 0 \\ 0.5 & 0.7 \end{bmatrix},$$

show that although $\|A\|_1 > 1$ *and* $\|A\|_\infty > 1$, *the matrix A is convergent.*

Here $\|A\|_1 = 1.3$, $\|A\|_\infty = 1.2$, and yet $A^r \to 0$ as $r \to \infty$, since the eigenvalues of A are 0.7 and 0.8.

Exercise

8. Show by calculation that

$$\begin{bmatrix} \frac{1}{2} & 0 & 0 \\ 1 & \frac{1}{2} & 0 \\ 0 & 1 & \frac{1}{2} \end{bmatrix}^r$$

$\to 0$ as $r \to \infty$.

Chapter 2

Parabolic Equations

2.1 Parabolic equations in one space dimension

Many problems in physics and engineering requiring numerical solution involve special cases of the linear parabolic partial differential equation

$$\sigma(x,t) \frac{\partial u}{\partial t} = \frac{\partial}{\partial x} \left(a(x,t) \frac{\partial u}{\partial x} \right) + b(x,t) \frac{\partial u}{\partial x} - c(x,t)u, \qquad (1)$$

which holds within some prescribed region R of the (x,t) space. Within this region, the functions σ, a are strictly positive and c is non-negative.

The region is usually one of three types illustrated in figure 1;

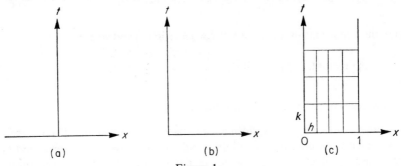

(a)   (b)   (c)

Figure 1

(a) the semi-infinite plane $[-\infty < x < +\infty] \times [t \geqslant 0]$. This leads to an initial value problem (Cauchy problem) consisting of equation (1) together with the initial condition

$$u = f(x) \quad (-\infty < x < +\infty) \text{ at } t = 0.$$

(b) the quarter plane $[x \geqslant 0] \times [t \geqslant 0]$. The relevant problem is the initial boundary value problem consisting of equation (1) together with the initial condition

$$u = f(x) \quad (0 \leqslant x < +\infty) \text{ at } t = 0$$

and the boundary condition at

$$x = 0, \quad t \geqslant 0$$

consisting of

$$\alpha_0(0,t)u + \alpha_1(0,t)\frac{\partial u}{\partial x} = \alpha_2(0,t),$$

where

$$\alpha_0(0,t) \geqslant 0, \quad \alpha_1(0,t) \leqslant 0, \quad \text{and} \quad \alpha_0 - \alpha_1 > 0.$$

[Some boundary information is also required at $x = \infty$, $t \geqslant 0$]

(c) the open rectangle $[0 \leqslant x \leqslant 1] \times [t \geqslant 0]$. Here the problem consists of equation (1) together with the initial condition

$$u = f(x) \quad (0 \leqslant x \leqslant 1) \text{ at } t = 0$$

and the boundary conditions consisting of

$$\alpha_0(0,t)u + \alpha_1(0,t)\frac{\partial u}{\partial x} = \alpha_2(0,t) \text{ at } x = 0, t \geqslant 0,$$

and

$$\beta_0(1,t)u + \beta_1(1,t)\frac{\partial u}{\partial x} = \beta_2(1,t) \text{ at } x = 1, t \geqslant 0,$$

where the conditions on the α's are the same as in (b) and where

$$\beta_0(1,t) \geqslant 0, \quad \beta_1(1,t) \geqslant 0, \quad \text{and} \quad \beta_0 - \beta_1 > 0.$$

2.2 Derivation of an exact difference formula

In order to obtain a finite difference replacement of equation (1), the region to be examined is covered by a rectilinear grid with sides parallel to the x- and t-axes, with h and k being the grid spacings in the x- and t-directions respectively (see figure 1c). The grid points (X,T) are given by $X = mh$, $T = nk$, where m and n are integers and $m = n = 0$ is the origin. The functions satisfying the difference and differential equations at the grid points $X = mh$, $T = nk$ are denoted by U_m^n and u_m^n respectively.

Returning to equation (1) in the form

$$\frac{\partial u}{\partial t} = L(x,t,D,D^2)u, \tag{2}$$

where the operator L is linear and $D = \partial/\partial x$, difference formulae involving two adjacent time levels and called *two level schemes* are obtained from the Taylor expansion

$$u(x, t+k) = \left(1 + k \frac{\partial}{\partial t} + \tfrac{1}{2}k^2 \frac{\partial^2}{\partial t^2} + \ldots \right) u(x,t)$$

$$= \exp \left(k \frac{\partial}{\partial t} \right) u(x,t).$$

If we now put $x = mh$, $t = nk$ and $u(mh,nk) = u_m^n$ then, provided L is independent of t,

$$u_m^{n+1} = \exp \left(k \frac{\partial}{\partial t} \right) u_m^n$$

$$= \exp(kL)u_m^n . \tag{3}$$

Now an exact formula* connecting D and δ_x, the central difference operator in the x-direction, is

$$D = \frac{2}{h} \sinh^{-1} \frac{\delta_x}{2} = \frac{1}{h} \left(\delta_x - \frac{1^2}{2^2 \cdot 3!} \delta_x^3 + \frac{1^2 \cdot 3^2}{2^4 \cdot 5!} \delta_x^5 \ldots \right), \tag{4}$$

where

$$\delta_x u_m^n = u_{m+1/2}^n - u_{m-1/2}^n ,$$

$$\delta_x^2 u_m^n = u_{m+1}^n - 2u_m^n + u_{m-1}^n ,$$

and so on. If equation (4) is used to eliminate D in terms of δ_x in equation (3), the exact difference replacement

$$u_m^{n+1} = \exp kL \left(mh,nk, \frac{2}{h} \sinh^{-1} \frac{\delta_x}{2}, \left(\frac{2}{h} \sinh^{-1} \frac{\delta_x}{2} \right)^2 \right) u_m^n \tag{5}$$

is obtained. All difference formulae in common use at present for solving equation (1) are approximations of equation (5).

There is no best method for obtaining approximating difference formulae, and as many different methods as possible will be demonstrated. The only requirement is that the formula, having been obtained, must pass certain tests of accuracy, consistency, stability and convergence, topics which will be discussed throughout this book.

---

* Many useful formulae connecting difference and differential operators can be found in F.B. Hildebrand, *Introduction to Numerical Analysis*, McGraw-Hill, 1956.

2.3 Explicit formulae

An explicit formula involves only one grid point at the advanced time level

$$t = (n + 1)k.$$

Consequently, only explicit formulae can be used to solve pure initial value problems (see figure 1). Explicit difference approximations are now derived for various forms of the heat conduction equation which are special cases of equation (1).

Case I   Constant coefficients

$$\frac{\partial u}{\partial t} = \frac{\partial^2 u}{\partial x^2}.$$ (6)

Here

$$L \equiv D^2,$$

and equation (3) becomes

$$u_m^{n+1} = \exp{(kD^2)}u_m^n,$$ (7)

where, from equation (4)

$$D^2 = \frac{1}{h^2} (\delta_x^2 - \tfrac{1}{12} \delta_x^4 + \tfrac{1}{90} \delta_x^6 \ldots).$$

Substitution of this value of D^2 in equation (7) followed by expansion leads to

$$u_m^{n+1} = [1 + r\delta_x^2 + \tfrac{1}{2}r(r - \tfrac{1}{6})\delta_x^4 + \tfrac{1}{6}r(r^2 - \tfrac{1}{2}r + \tfrac{1}{15})\delta_x^6 \ldots]u_m^n,$$ (8)

where $r = k/h^2$ is the mesh ratio. From equation (8), if we retain only second order central differences, the forward difference formula

$$U_m^{n+1} = (1 + r\delta_x^2)U_m^n,$$

is obtained which, on substitution for δ_x^2, leads to

$$U_m^{n+1} = (1 - 2r)U_m^n + r(U_{m+1}^n + U_{m-1}^n),$$ (9)

where U_m^n is an approximation to u_m^n. If fourth order central differences are retained, equation (8) gives rise to the difference formula

$$U_m^{n+1} = \tfrac{1}{2}(2 - 5r + 6r^2)U_m^n + \tfrac{2}{3}r(2 - 3r)(U_{m+1}^n + U_{m-1}^n) -$$

$$\tfrac{1}{12}r(1 - 6r)(U_{m+2}^n + U_{m-2}^n),\tag{10}$$

where again U_m^n is an approximation to u_m^n.

It is instructive at this stage to establish a criterion for the local accuracy of a finite difference formula. In particular this will be done for the basic explicit formula (9). Introduce the difference between the exact solutions of the differential and difference equations at the grid point (mh, nk) as

$$z_m^n = u_m^n - U_m^n.\tag{11}$$

Using Taylor's theorem

$$u_m^{n+1} = u_m^n + k\left(\frac{\partial u}{\partial t}\right)_m^n + \tfrac{1}{2}k^2\left(\frac{\partial^2 u}{\partial t^2}\right)_m^n + \dots,$$

$$u_{m+1}^n = u_m^n + h\left(\frac{\partial u}{\partial x}\right)_m^n + \tfrac{1}{2}h^2\left(\frac{\partial^2 u}{\partial x^2}\right)_m^n + \tfrac{1}{6}h^3\left(\frac{\partial^3 u}{\partial x^3}\right)_m^n + \tfrac{1}{24}h^4\left(\frac{\partial^4 u}{\partial x^4}\right)_m^n + \dots$$

and

$$u_{m-1}^n = u_m^n - h\left(\frac{\partial u}{\partial x}\right)_m^n + \tfrac{1}{2}h^2\left(\frac{\partial^2 u}{\partial x^2}\right)_m^n - \tfrac{1}{6}h^3\left(\frac{\partial^3 u}{\partial x^3}\right)_m^n + \tfrac{1}{24}h^4\left(\frac{\partial^4 u}{\partial x^4}\right)_m^n \dots,$$

and so

$$u_m^{n+1} - (1 - 2r)u_m^n - r(u_{m+1}^n + u_{m-1}^n) =$$

$$k\left(\frac{\partial u}{\partial t} - \frac{\partial^2 u}{\partial x^2}\right)_m^n + \tfrac{1}{2}k^2\left(\frac{\partial^2 u}{\partial t^2} - \frac{1}{6r}\frac{\partial^4 u}{\partial x^4}\right)_m^n + \dots.\tag{12}$$

From equations (6), (9), (11) and (12), the result

$$z_m^{n+1} = (1 - 2r)z_m^n + r(z_{m+1}^n + z_{m-1}^n) + \tfrac{1}{2}k^2\left(\frac{\partial^2 u}{\partial t^2} - \frac{1}{6r}\frac{\partial^4 u}{\partial x^4}\right)_m^n + \dots\tag{13}$$

is obtained. The quantity

$$\tfrac{1}{2}k^2\left(\frac{\partial^2 u}{\partial t^2} - \frac{1}{6r}\frac{\partial^4 u}{\partial x^4}\right)_m^n + \dots$$

is defined as the local truncation error* of formula (9), and the *principal part* of

---

* The *local order of accuracy* (often abbreviated to local accuracy) of formula (9) is $0(k + h^2)$. This should not be confused with the *global accuracy* of a difference formula, which is a measure of the accuracy of the difference formula *all over* the region under consideration and is a very difficult quantity to estimate, well beyond the scope of this book.

B

the truncation error is

$$\tfrac{1}{2}k^2\left(\frac{\partial^2 u}{\partial t^2} - \frac{1}{6r}\frac{\partial^4 u}{\partial x^4}\right)^n_m.$$

It is assumed that all the derivatives of u with respect to x and t are bounded. Alternatively, if we use the order notation,

$$F(x) = 0(G(x))$$

means that

$$|F(x)| \leqslant AG(x) \quad (A, \text{a positive constant})$$

as x tends to a prescribed limit, and equation (13) can be written as

$$z_m^{n+1} = (1 - 2r)z_m^n + r(z_{m+1}^n + z_{m-1}^n) + 0(k^2 + kh^2).$$

It should be pointed out that, if $u \in C^{m,n}$ means that the derivatives of u with respect to x and t are continuous up to orders m and n respectively, then, in the present example, we require $u \in C^{4,2}$ in order to define the principal part of the truncation error.

Exercise

1. Find the principal part of the truncation error of the difference formula (10).

Case II   Coefficients depending on x

$$\frac{\partial u}{\partial t} = a(x)\,\frac{\partial^2 u}{\partial x^2}, \tag{14}$$

where $a(x) \neq 0$ for all x. This time

$$L \equiv aD^2,$$

and equation (3) becomes

$$\begin{aligned}
u_m^{n+1} &= \exp(kaD^2)u_m^n \\
&= [1 + kaD^2 + \tfrac{1}{2}k^2aD^2(aD^2) + \dots]u_m^n \\
&= [1 + kaD^2 + \tfrac{1}{2}k^2a(a''D^2 + 2a'D^3 + aD^4) + \dots]u_m^n,
\end{aligned}$$

where $'$ denotes differentiation with respect to x. The differential operators D^2, D^3, $D^4, \dots$ are now replaced by difference operators using equation (4). The presence of $a(x)$ has considerably complicated matters and the only explicit difference formula in common use for this case is

$$U_m^{n+1} = (1 - 2ra)U_m^n + ra(U_{m+1}^n + U_{m-1}^n), \tag{15}$$

where a is evaluated at $x = mh$.

Exercise

2. Derive formula (15) indicating where truncation has occurred.

Case III *The self-adjoint case*

$$\frac{\partial u}{\partial t} = \frac{\partial}{\partial x}\left(a(x)\frac{\partial u}{\partial x}\right), \tag{16}$$

where $a(x) \neq 0$ for all x. Differentiation of the right-hand side gives

$$L \equiv aD^2 + a'D,$$

and so

$$u_m^{n+1} = \exp k(aD^2 + a'D)u_m^n$$

$$= [1 + k(aD^2 + a'D) + \dots]u_m^n.$$

If terms of order k^2 and above are neglected and $D^2 u_m^n$ and Du_m^n are replaced by $(1/h^2)\delta^2 u_m^n$ and $(1/2h)(u_{m+1}^n - u_{m-1}^n)$ respectively, the formula

$$U_m^{n+1} = (1 - 2ra)U_m^n + r(a + \tfrac{1}{2}ha')U_{m+1}^n + r(a - \tfrac{1}{2}ha')U_{m-1}^n \tag{17}$$

is obtained.

Example 1 *Show that the local truncation error of formula (17) is $O(k^2 + kh^2)$.*

Formula (17) can be rewritten as

$$U_m^{n+1} = (1 - 2ra)U_m^n + ra(U_{m+1}^n + U_{m-1}^n) + \tfrac{1}{2}hra'(U_{m+1}^n - U_{m-1}^n).$$

If Taylor's theorem is used with a remainder term (see any standard work on the calculus),

$$u_{m+1}^n = u_m^n + h\left(\frac{\partial u}{\partial x}\right)_m^n + \tfrac{1}{2}h^2\left(\frac{\partial^2 u}{\partial x^2}\right)_m^n + \tfrac{1}{6}h^3\left(\frac{\partial^3 u}{\partial x^3}\right)_m^n + \tfrac{1}{24}h^4\left(\frac{\partial^4 u}{\partial x^4}\right)_{m+\beta}^n,$$

$$u^n_{m-1} = u^u_m - h\left(\frac{\partial u}{\partial x}\right)^n_m + \tfrac{1}{2}h^2\left(\frac{\partial^2 u}{\partial x^2}\right)^n_m - \tfrac{1}{6}h^3\left(\frac{\partial^3 u}{\partial x^3}\right)^n_m + \tfrac{1}{24}h^4\left(\frac{\partial^4 u}{\partial x^4}\right)^n_{m-\gamma},$$

$$u^{n+1}_m = u^n_{m.} + k\left(\frac{\partial u}{\partial t}\right)^n_m + \tfrac{1}{2}k^2\left(\frac{\partial^2 u}{\partial t^2}\right)^{n+\alpha}_m,$$

for $0 < \alpha, \beta, \gamma < 1$, then

$$u^{n+1}_m = (1 - 2ra)u^n_m + ra(u^n_{m+1} + u^n_{m-1}) + \tfrac{1}{2}hra'(u^n_{m+1} - u^n_{m-1}) +$$

$$k\left(\frac{\partial u}{\partial t} - \frac{\partial}{\partial x}\left(a\frac{\partial u}{\partial x}\right)\right)^n_m + \tfrac{1}{2}k^2\left\{\left(\frac{\partial^2 u}{\partial t^2}\right)^{n+\alpha}_m - \frac{1}{6r}\left[\tfrac{1}{2}a\left(\left(\frac{\partial^4 u}{\partial x^4}\right)^n_{m+\beta} +\right.\right.\right.$$

$$\left.\left.\left.\left(\frac{\partial^4 u}{\partial x^4}\right)^n_{m-\gamma}\right) + 2a'\left(\frac{\partial^3 u}{\partial x^3}\right)^n_m\right] - \frac{h}{24r}a'\left[\left(\frac{\partial^4 u}{\partial x^4}\right)^n_{m+\beta} - \left(\frac{\partial^4 u}{\partial x^4}\right)^n_{m-\gamma}\right]\right\}$$

Since u satisfies the original differential equation (16), the term in k disappears and so the desired result is obtained.

Alternatively, difference approximations to equation (16) can be obtained without differentiating the right-hand side and destroying the self-adjoint nature of the operator. The following method is due to Tikhonnov and Samarskii (1961). Write equation (16) in the conservation form

$$\frac{\partial u}{\partial t} + \frac{\partial w}{\partial x} = 0, \tag{18}$$

where

$$w = -a(x)\frac{\partial u}{\partial x}.$$

Rewrite the latter formula as

$$\frac{w}{a(x)} = -\frac{\partial u}{\partial x},$$

and integrate with respect to x over the interval $[(m-1)h, mh]$ of the x-axis. If it is *assumed* that $w = w_{m-1/2}$ over this interval, the result

$$w_{m-1/2}\int^{mh}_{(m-1)h} \frac{dx}{a(x)} = u_{m-1} - u_m$$

is obtained. Similarly, if the interval $[mh, (m + 1)h]$ is considered, we get

$$w_{m+1/2} \int_{mh}^{(m+1)h} \frac{dx}{a(x)} = u_m - u_{m+1}.$$

Now, from equation (18)

$$\frac{\partial u}{\partial t} = -\frac{\partial w}{\partial x}$$

$$= -\frac{1}{h}\,\delta_x w + 0(h^2) \qquad \text{(from equation 4)}$$

$$= \frac{1}{h}\,(w_{m-1/2} - w_{m+1/2}) + 0(h^2) \qquad \text{(from equation 4)}$$

$$= B_m(u_{m+1} - u_m) - A_m(u_m - u_{m-1}) + 0(h^2), \qquad (19)$$

where

$$B_m = \frac{1}{h}\left[\int_{mh}^{(m+1)h} \frac{dx}{a(x)}\right]^{-1}, \quad A_m = \frac{1}{h}\left[\int_{(m-1)h}^{mh} \frac{dx}{a(x)}\right]^{-1}.$$

Since $B_m = A_{m+1}$, equation (19) can be written as

$$\left(\frac{\partial u}{\partial t}\right)_m^n = A_{m+1}u_{m+1} - (A_{m+1} + A_m)u_m + A_m u_{m-1} + 0(h^2), \qquad (20)$$

and since

$$\left(\frac{\partial u}{\partial t}\right)_m^n = \frac{1}{k}(u_m^{n+1} - u_m^n) + 0(k),$$

the formula

$$U_m^{n+1} = [1 - k(A_{m+1} + A_m)]\,U_m^n + k[A_{m+1} U_{m+1}^n + A_m U_{m-1}^n] \qquad (21)$$

is obtained.

A more standard method of approximating equation (16) is to consider the central difference approximation to the self-adjoint operator, namely

$$\frac{1}{h}\,\delta_x(a_m\delta_x)u_m = \frac{1}{h}\,\delta_x[a_m(u_{m+1/2} - u_{m-1/2})]$$

$$= \frac{1}{h^2} \left[a_{m+1/2} u_{m+1} - a_{m-1/2} u_m - a_{m+1/2} u_m + a_{m-1/2} u_{m-1} \right]$$

$$= \frac{1}{h^2} \left[a_{m+1/2} (u_{m+1} - u_m) - a_{m-1/2} (u_m - u_{m-1}) \right].$$

This leads to the explicit difference formula

$$U_m^{n+1} = \left[1 - r(a_{m+1/2} + a_{m-1/2}) \right] U_m^n + r \left[a_{m+1/2} U_{m+1}^n + a_{m-1/2} U_{m-1}^n \right]. \quad (22)$$

Although formulae (21) and (22) are very similar in form, the greater dependence of formula (21) on the nature of $a(x)$ makes it superior in many cases.

Exercise

3. If

$$a(x) = x^{-s}, (s \geqslant 0)$$

show that formulae (21) and (22) are equivalent for $s = 0, 1$.
[Hint: formulae (21) and (22) are equivalent if $h^2 A_{m+1} = a_{m+1/2}$.]

In Cases II and III, if $a \equiv a(x, t)$, formulae (15), (17), (21) and (22) still hold with the same order of accuracy.

Case IV Significant first space derivative

$$\frac{\partial u}{\partial t} = \frac{\partial^2 u}{\partial x^2} - b \frac{\partial u}{\partial x} \quad (23)$$

with b a positive constant which can be large. This is a most important case as it is a simplified model of many convective diffusion and viscous flow problems. If we put $a = 1$ and $a' = -b$ in (17) we get

$$U_m^{n+1} = (1 - 2r)U_m^n + r(1 - L)U_{m+1}^n + r(1 + L)U_{m-1}^n \quad (24)$$

where $L = \frac{1}{2} hb > 0$. The solution oscillates for $L > 1$ (Siemieniuch and Gladwell (1978)) and so for large values of b it is customary to replace central differences for the first space derivative by *backward* or *upstream* differences. Although these reduce the local accuracy of the difference scheme from second to first order, they get rid of the oscillations. The difference formula where D is replaced by $\frac{1}{h} \nabla_x$ with

$$\nabla_x U_m^n = U_m^n - U_{m-1}^n$$

is given by

$$U_m^{u+1} = \left[1 - 2r(1 + L) \right] U_m^n + r U_{m+1}^n + r(1 - 2L) U_{m-1}^n. \quad (25)$$

Further consideration of this case with a significant first space derivative is given in Chapters 5 and 6.

Case V Cylindrical polar co-ordinates, (ρ, θ)

$$\frac{\partial u}{\partial t} = \frac{\partial^2 u}{\partial \rho^2} + \frac{1}{\rho} \frac{\partial u}{\partial \rho} \tag{26}$$

in the axially symmetric region $(0 \leqslant \rho \leqslant 1) \times (t \geqslant 0)$ with the boundary condition $\partial u/\partial \rho = 0$ at $\rho = 0$ for all $t \geqslant 0$. It is a feature of numerical solutions of this equation using finite differences on a rectangular grid that the solution is less accurate in the vicinity of the cylinder axis $(\rho = 0)$ than in the remainder of the field. This is due to the term $1/\rho(\partial u/\partial \rho)$ which is difficult to represent adequately by finite differences when ρ is small. On the axis of course, where $\rho = 0$, the term becomes $\partial^2 u/\partial \rho^2$. If the variable ρ is transformed according to the relation

$$\rho = 2x^{1/2},$$

the original equation becomes

$$\frac{\partial u}{\partial t} = \frac{\partial u}{\partial x} + x \frac{\partial^2 u}{\partial x^2} \quad (0 \leqslant x \leqslant \frac{1}{4}). \tag{27}$$

In the (x, t) plane we construct a rectangular grid with equal spacing in the t-direction given by $t = nk$ $(n = 0, 1, 2, \ldots)$ and unequal spacing in the x-direction given by $x = m^2 h$ $(m = 0, 1, 2, \ldots)$, where h, k are constants. The latter is consistent with equal spacing in the original ρ direction. An explicit difference replacement referred to this grid is

$$U_m^{n+1} = \left[1 - \frac{2p(m^2+p-1)}{4m^2-1} \right] U_m^n + \frac{p(2m^2+2m+2p-1)}{4m(2m-1)} U_{m+1}^n +$$

$$+ \frac{p(2m^2-2m+2p-1)}{4m(2m-1)} U_{m-1}^n \quad (m = 1, 2, \ldots) \tag{28}$$

where $p = k/h$, and

$$U_0^{n+1} = \frac{1}{4}(4 - 5p + 2p^2)U_0^n - \frac{2}{3}p(p-2)U_1^n + \frac{1}{12}p(2p-1)U_2^n$$

for points on the axis. These formulae are derived in Mitchell and Pearce (1963) along with an assessment of the truncation error and a stability condition. An increased accuracy difference scheme for the heat conduction equation in cylindrical polar co-ordinates is given in Polak (1974).

2.4 Implicit formulae

An implicit formula involves more than one grid point at the advanced time level $t = (n + 1)k$. Some important implicit formulae are now derived for particular cases of equation (1). These formulae can often be obtained from equation (3) written in the central form

$$\exp\left(-\tfrac{1}{2}kL\right)u_m^{n+1} = \exp\left(\tfrac{1}{2}kL\right)u_m^n. \tag{29}$$

Case I Constant coefficients

$$\frac{\partial u}{\partial t} = \frac{\partial^2 u}{\partial x^2}.$$

Here $L \equiv D^2$, and so equation (29) becomes

$$\exp\left(-\tfrac{1}{2}kD^2\right)u_m^{n+1} = \exp\left(\tfrac{1}{2}kD^2\right)u_m^n. \tag{30}$$

Correct to second differences

$$D^2 = \frac{1}{h^2}\,\delta_x^2,$$

and substitution into equation (30) followed by expansion leads to the central difference formula

$$(1 - \tfrac{1}{2}r\delta_x^2)U_m^{n+1} = (1 + \tfrac{1}{2}r\delta_x^2)U_m^n, \tag{31}$$

with a principal truncation error $O(k^3 + kh^2)$. This is the *Crank–Nicolson formula* and may be written in the form

$$(1 + r)U_m^{n+1} - \tfrac{1}{2}r(U_{m+1}^{n+1} + U_{m-1}^{n+1}) = (1 - r)U_m^n + \tfrac{1}{2}r(U_{m+1}^n + U_{m-1}^n).$$

The formula of maximum accuracy based on the same six grid points as the Crank-Nicolson formula is obtained by substituting into equation (30)

$$D^2 = \frac{1}{h^2}\,\delta_x^2\left(1 + \tfrac{1}{12}\,\delta_x^2\right)^{-1}$$

a result which is correct to fourth differences, and expanding equation (30) to give the Douglas formula

$$[1 - \tfrac{1}{2}(r - \tfrac{1}{6})\delta_x^2]U_m^{n+1} = [1 + \tfrac{1}{2}(r + \tfrac{1}{6})\delta_x^2]U_m^n, \tag{32}$$

with a local truncation error that is $O(k^3 + kh^4)$.

Example 2 Solve

$$\frac{\partial u}{\partial t} = \frac{\partial^2 u}{\partial x^2}$$

subject to the initial condition $u = \sin x (0 \leqslant x \leqslant \pi)$ *at* $t = 0$ *and the boundary conditions* $u = 0$ *at* $x = 0, \pi (t \geqslant 0)$, *using*

(i)   the Crank–Nicolson formula (31).
(ii) the Douglas formula (32).

In each case take $h = \pi/20$ and $r = 1/\sqrt{20}$. (This is the value of r for which the Douglas formula has minimum truncation error). Use the method of solution of tridiagonal systems (given in section 2.5) to solve the set of equations at each time step. This example should be evaluated on a high speed computer.
    The theoretical solution (T.S.) for the differential system is given by

$$u = e^{-t} \sin x,$$

and the errors for the Crank–Nicolson (E_{CN}) and the Douglas (E_D) schemes are given in table 1 at $x = \pi/2$ after a number of time steps (P).

Case II Variable coefficients

$$\frac{\partial u}{\partial t} = a(x, t) \frac{\partial^2 u}{\partial x^2}. \tag{33}$$

It is a simple exercise to show by Taylor expansions that

$$\frac{1}{2h^2}\delta_x^2(u_m^{n+1} + u_m^n) = \left(\frac{\partial^2 u}{\partial x^2}\right)_m^{n+1/2} + \tfrac{1}{12}h^2\left(\frac{\partial^4 u}{\partial x^4}\right)_m^{n+1/2} + 0(h^4 + k^2)$$

$$= \left(\frac{1}{a}\frac{\partial u}{\partial t}\right)_m^{n+1/2} + \tfrac{1}{12}h^2\frac{\partial^2}{\partial x^2}\left(\frac{1}{a}\frac{\partial u}{\partial t}\right)_m^{n+1/2} + 0(h^4 + k^2)$$

(from equation 33) $\tag{34}$

Table 1

P	T.S.	E_{CN}	E_D
1	0.994,497,915,630	0.000,011	−0.000,000,000,026
2	0.989,026,104,192	0.000,022	−0.000,000,000,051
4	0.978,172,634,773	0.000,040	−0.000,000,000,101
8	0.956,821,703,419	0.000,079	−0.000,000,000,198
16	0.915,507,772,134	0.000,151	−0.000,000,000,379
80	0.643,146,895,793	0.000,531	−0.000,000,001,331
160	0.413,637,929,568	0.000,683	−0.000,000,001.712
320	0.171,096,336,778	0.000,565	−0.000,000,001,417
640	0.029,273,956,459	0.000,194	−0.000,000,000,485
800	0.012,108,818,740	0.000,100	−0.000,000,000,251

30

It can also be shown that

$$\frac{\partial^2}{\partial x^2}\left(\frac{1}{a}\frac{\partial u}{\partial t}\right)_m^{n+1/2} = \frac{1}{h^2 k}\delta_x^2\left[\frac{1}{a_m^{n+1/2}}(u_m^{n+1} - u_m^n)\right] + O(h^2 + k^2),$$

and substitution in equation (34) gives

$$\frac{1}{2h^2}\delta_x^2(u_m^{n+1} + u_m^n) = \frac{1}{a_m^{n+1/2}k}(u_m^{n+1} - u_m^n) +$$

$$\frac{1}{12r}\frac{1}{h^2}\delta_x^2\left[\frac{1}{a_m^{n+1/2}}(u_m^{n+1} - u_m^n)\right] + O(h^4 + k^2)$$

which leads to the difference equation

$$\frac{1}{ra_m^{n+1/2}}(U_m^{n+1} - U_m^n) = \tfrac{1}{2}\delta_x^2\left[1 - \frac{1}{6ra_m^{n+1/2}}\right]U_m^{n+1} + \tfrac{1}{2}\delta_x^2\left[1 + \frac{1}{6ra_m^{n+1/2}}\right]U_m^n,$$

(35)

with a local truncation error that is $O(h^6) = O(k^3)$, when r is kept fixed. Formula (35) can be rewritten in the more conveneient form

$$[1 + \tfrac{1}{12}a_m^{n+1/2}\delta_x^2(a_m^{n+1/2})^{-1} - \tfrac{1}{2}ra_m^{n+1/2}\delta_x^2]U_m^{n+1} =$$

$$[1 + \tfrac{1}{12}a_m^{n+1/2}\delta_x^2(a_m^{n+1/2})^{-1} + \tfrac{1}{2}ra_m^{n+1/2}\delta_x^2]U_m^n. \quad (35a)$$

Exercise

4. Show by Taylor expansions that

(i) $\dfrac{1}{2h^2}\delta_x^2(u_m^{n+1} + u_m^n) = \left(\dfrac{\partial^2 u}{\partial x^2}\right)_m^{n+1/2} + \tfrac{1}{12}h^2\left(\dfrac{\partial^4 u}{\partial x^4}\right)_m^{n+1/2} + O(h^4 + k^2)$

(ii) $\dfrac{1}{h^2 k}\delta_x^2\left[\dfrac{1}{a_m^{n+1/2}}(u_m^{n+1} - u_m^n)\right] = \dfrac{\partial^2}{\partial x^2}\left(\dfrac{1}{a}\dfrac{\partial u}{\partial t}\right)_m^{n+1/2} + O(h^2 + k^2).$

Case III

$$\frac{\partial u}{\partial t} = \frac{\partial}{\partial x}\left(a(x,t)\frac{\partial u}{\partial x}\right) + b(x,t)\frac{\partial u}{\partial x} + c(x,t)u. \quad (36)$$

Again we consider equation (36) to be located at $x = mh$, $t = (n + \tfrac{1}{2})k$, and so

$$\frac{1}{k}(u_m^{n+1} - u_m^n) = \tfrac{1}{2}\left[\frac{\partial}{\partial x}\left(a\frac{\partial u}{\partial x}\right) + b\frac{\partial u}{\partial x} + cu\right]_m^{n+1} +$$

$$\frac{1}{2}\left[\frac{\partial}{\partial x}\left(a\frac{\partial u}{\partial x}\right)+b\frac{\partial u}{\partial x}+cu\right]_m^n+0(k^2).\qquad(37)$$

Now at any fixed time

$$\frac{\partial}{\partial x}\left(a\frac{\partial u}{\partial x}\right)=A_{m+1}u_{m+1}-(A_{m+1}+A_m)u_m+A_mu_{m-1}+0(h^2)$$

(from equation 20),

where

$$A_m=\frac{1}{h}\left[\int_{(m-1)h}^{mh}\frac{dx}{a(x)}\right]^{-1},$$

and

$$b\frac{\partial u}{\partial x}=b_m\frac{1}{2h}(u_{m+1}-u_{m-1})+0(h^2).$$

Substitution of these values into equation (37) leads to the implicit difference scheme

$$[1+\tfrac{1}{2}k(A_{m+1}^{n+1}+A_m^{n+1})-\tfrac{1}{2}kc_m^{n+1}]U_m^{n+1}-$$

$$\tfrac{1}{2}k\left[\left(A_{m+1}^{n+1}+\frac{1}{2h}b_m^{n+1}\right)U_{m+1}^{n+1}+\left(A_m^{n+1}-\frac{1}{2h}b_m^{n+1}\right)U_{m-1}^{n+1}\right]=$$

$$[1-\tfrac{1}{2}k(A_{m+1}^n+A_m^n)+\tfrac{1}{2}kc_m^n]U_m^n+$$

$$\tfrac{1}{2}k\left[\left(A_{m+1}^n+\frac{1}{2h}b_m^n\right)U_{m+1}^n+\left(A_m^n-\frac{1}{2h}b_m^n\right)U_{m-1}^n\right].\qquad(38)$$

If b is large in modulus, we put

$$b\frac{\partial u}{\partial x}=b_m\frac{1}{h}(U_m-U_{m-1})+0(h)\qquad\qquad b<0$$

$$=b_m\frac{1}{h}(U_{m+1}-U_m)+0(h)\qquad\qquad b>0$$

and alter equation (38) accordingly.

Another class of implicit formulae for approximating particular cases of equation (1) involves only one grid point at the first time level $t=nk$. These so-called backward difference formulae are derived from equation (3) written in the form

$$\exp\left(-kL\right) u_m^{n+1} = u_m^n. \tag{39}$$

For example, when

$$L \equiv D^2 = \frac{1}{h^2}\delta_x^2 + 0(h^2),$$

expansion of equation (39) leads to the backward difference implicit formula

$$(1 - r\delta_x^2)U_m^{n+1} = U_m^n. \tag{40}$$

Formulae (3), (29) and (39) are the special cases $\theta = 0$, ½, and 1 respectively of the general θ formula

$$\exp\left(-\theta kL\right)u_m^{n+1} = \exp\left((1 - \theta)kL\right)u_m^n, \qquad 0 \leqslant \theta \leqslant 1.$$

2.5 Solution of tridiagonal systems

The implicit difference formulae introduced in the previous section have all involved three unknown values of U at the advanced time level $t = (n + 1)k$. It is not difficult to see that these implicit formulae can only be used to solve initial boundary value problems of the type illustrated in figure 1(c) where U is given on $x = 0$, $1(t > 0)$. If $Mh = 1$, the system of linear algebraic equations arising from the implicit difference formulae which must be solved at each time step is a special case of the tridiagonal system,

$$-\alpha_m U_{m-1} + \beta_m U_m - \gamma_m U_{m+1} = \delta_m,$$

for $1 \leqslant m \leqslant M - 1$, where U_0 and U_M are known from the boundary conditions. If

$$\alpha_m > 0, \quad \beta_m > 0, \quad \gamma_m > 0,$$

and

$$\beta_m \geqslant (\alpha_m + \gamma_m),$$

for $1 \leqslant m \leqslant M - 1$, a highly efficient method suitable for automatic computation is available for solving the tridiagonal system. The name tridiagonal system arises from the fact that, if the system of equation is written in matrix form, namely,

$$
\begin{bmatrix}
\beta_1 & -\gamma_1 & & & & \bigcirc \\
-\alpha_2 & \beta_2 & -\gamma_2 & & & \\
 & \cdot & \cdot & \cdot & & \\
 & & \cdot & \cdot & \cdot & \\
 & & & \cdot & \cdot & -\gamma_{M-2} \\
\bigcirc & & & & -\alpha_{M-1} & \beta_{M-1}
\end{bmatrix}
\begin{bmatrix}
U_1 \\ U_2 \\ \\ \\ U_{M-2} \\ U_{M-1}
\end{bmatrix}
=
\begin{bmatrix}
\delta_1 + \alpha_1 U_0 \\
\delta_2 \\ \\ \\
\delta_{M-2} \\
\delta_{M-1} + \gamma_{M-1} U_M
\end{bmatrix}
$$

the matrix is tridiagonal.

The method is obtained in the following manner. Consider the difference relation

$$U_m = w_m U_{m+1} + g_m,$$

for $0 \leqslant m \leqslant M - 1$, from which it follows that

$$U_{m-1} = w_{m-1} U_m + g_{m-1}.$$

If this is used to eliminate U_{m-1} from the original difference formula defining the tridiagonal system, the result

$$U_m = \frac{\gamma_m}{\beta_m - \alpha_m w_{m-1}} U_{m+1} + \frac{\delta_m + \alpha_m g_{m-1}}{\beta_m - \alpha_m w_{m-1}}$$

is obtained, and so

$$w_m = \frac{\gamma_m}{\beta_m - \alpha_m w_{m-1}}, \qquad g_m = \frac{\delta_m + \alpha_m g_{m-1}}{\beta_m - \alpha_m w_{m-1}}.$$

If $U_0 = 0$, then $w_0 = g_0 = 0$, in order that the difference relation

$$U_0 = w_0 U_1 + g_0$$

holds for any U_1. The remaining w_m, g_m $(m = 1, 2, \ldots, M - 1)$ can now be calculated from

$$w_1 = \frac{\gamma_1}{\beta_1} \qquad g_1 = \frac{\delta_1}{\beta_1}$$

$$w_2 = \frac{\gamma_2}{\beta_2 - \alpha_2 w_1} \qquad g_2 = \frac{\delta_2 + \alpha_2 g_1}{\beta_2 - \alpha_2 w_1}$$

$$w_{M-1} = \frac{\gamma_{M-1}}{\beta_{M-1} - \alpha_{M-1} w_{M-2}} \qquad g_{M-1} = \frac{\delta_{M-1} + \alpha_{M-1} g_{M-2}}{\beta_{M-1} - \alpha_{M-1} w_{M-2}}.$$

If $U_M = Y$, then $U_1, U_2, \ldots, U_{M-1}$ are calculated from

$$U_{M-1} = w_{M-1}\, Y + g_{M-1}$$

$$U_{M-2} = w_{M-2}\, U_{M-1} + g_{M-2}$$

$$\cdot$$
$$\cdot$$
$$\cdot$$

$$U_1 = w_1 U_2 + g_1.$$

In using this method, substantial errors will appear in the calculated values of U_1, $U_2, \ldots, U_{M-1}$ unless

$$|w_m| \leqslant 1 \quad (m = 1, 2, \ldots, M-1).$$

Now

$$w_1 = \frac{\gamma_1}{\beta_1} \leqslant 1,$$

$$w_2 = \frac{\gamma_2}{\beta_2 - \alpha_2 w_1} \leqslant \frac{\gamma_2}{\beta_2 - \alpha_2} \leqslant 1,$$

and so on, since $\alpha_m > 0$, $\beta_m > 0$, $\gamma_m > 0$, $\beta_m \geqslant (\alpha_m + \gamma_m)$ for $1 \leqslant m \leqslant M-1$. This leads to

$$0 < w_m \leqslant 1 \quad (1 \leqslant m \leqslant M-1).$$

While this method is equivalent to Gaussian elimination, it avoids the error growth associated with the back substitution in the elimination method, and also minimizes the storage in the machine computation. In fact, with many machines the solution of a problem by this method is more efficient than the solution using an explicit difference scheme where the solution is calculated directly. If the boundary conditions are not given by $U_0 = 0$, $U_M = Y$, this method can still be used with very slight modifications. Further details of the method are available on page 395 of Todd (1962).

Example 3 Illustrate the method described for solving a tridiagonal system with reference to the Crank–Nicolson method to solve an initial boundary value problem of the type illustrated in figure 1(c) involving the differential equation

$$\frac{\partial u}{\partial t} = \frac{\partial^2 u}{\partial x^2}.$$

The Crank—Nicolson formula (31) is

$$(1 + r)U_m^{n+1} - \tfrac{1}{2}r(U_{m+1}^{n+1} + U_{m-1}^{n+1}) = (1 - r)U_m^n + \tfrac{1}{2}r(U_{m+1}^n + U_{m-1}^n).$$

If we choose $M = 8(h = \tfrac{1}{8})$, the system of equations to be solved is

$$(1 + r)U_1^{n+1} - \tfrac{1}{2}rU_2^{n+1} = \tfrac{1}{2}rU_0^n + (1 - r)U_1^n + \tfrac{1}{2}rU_2^n + \tfrac{1}{2}rU_0^{n+1}$$

$$-\tfrac{1}{2}rU_1^{n+1} + (1 + r)U_2^{n+1} - \tfrac{1}{2}rU_3^{n+1} = \tfrac{1}{2}rU_1^n + (1 - r)U_2^n + \tfrac{1}{2}rU_3^n$$

$$\vdots$$

$$-\tfrac{1}{2}rU_5^{n+1} + (1 + r)U_6^{n+1} - \tfrac{1}{2}rU_7^{n+1} = \tfrac{1}{2}rU_5^n + (1 - r)U_6^n + \tfrac{1}{2}rU_7^n$$

$$- \tfrac{1}{2}rU_6^{n+1} + (1 + r)U_7^{n+1} = \tfrac{1}{2}rU_6^n + (1 - r)U_7^n + \tfrac{1}{2}rU_8^n + \tfrac{1}{2}\,rU_8^{n+1}.$$

Since $U_1^n, U_2^n, \ldots, U_7^n$, are values calculated previously, and U_0^n, U_8^n, U_0^{n+1}, U_8^{n+1} are given boundary values, the right-hand side of the system of equations is known, and hence the values $\delta_m (m = 1, 2, \ldots, 7)$ of the general tridiagonal system are known. In addition

$$\begin{aligned}
\alpha_m &= + \tfrac{1}{2}r > 0 \quad (m = 1, 2, \ldots, 7), \\
\beta_m &= 1 + r > 0 \quad (m = 1, 2, \ldots, 7), \\
\gamma_m &= + \tfrac{1}{2}r > 0 \quad (m = 1, 2, \ldots, 7), \\
\beta_m &> (\alpha_m + \gamma_m) \ (m = 1, 2, \ldots, 7),
\end{aligned}$$

and so the conditions of the method are satisfied for all grid ratios r. The values of $w_0, w_1, \ldots, w_7$ and $g_0, g_1, \ldots g_7$ are now easily calculated from the known values $\alpha_m, \beta_m, \gamma_m, \delta_m, (m = 1, 2, \ldots, 7)$ and hence $U_7^{n+1}, U_6^{n+1}, \ldots, U_1^{n+1}$ are obtained.

2.6 Convergence

So far, several finite difference schemes have been derived for solving particular cases of equation (1). These schemes which are of varying degrees of accuracy are only of use as a means of solving equation (1) if they are convergent and stable.

The problem of convergence of a finite difference method for solving equation (1) consists of finding the conditions under which

$$u(X,T) - U(X,T),$$

the difference between the theoretical solutions of the differential and difference equations at a *fixed* point (X,T), tend to zero uniformly, as the net is refined in such a way that $h, k \to 0$ and $m, n \to \infty$, with $mh(= X)$ and $nk(= T)$ remaining fixed. The fixed point (X,T) is anywhere within the region R under consideration, and it is sometimes necessary in the convergence analysis to assume that h, k do not tend to zero independently but according to some relationship like

36

$$k = rh^2,$$ (41)

where r is a constant.

Before carrying out a convergence analysis, it is instructive to examine the domain of dependence of a fixed point (X,T) as $h,k \to 0$ according to the relationship (41), when an explicit formula, such as (9), is used for calculation. The domain of dependence includes all grid points which contribute to the calculated value at the point (X,T). In this case (see figure 2), it is a triangle with a semi-apex angle at (X,T) given by

$$\theta = \tan^{-1} \frac{h}{k} = \tan^{-1} \frac{1}{rh},$$

with its base on the x-axis. As $h \to 0$ with r fixed, $\theta \to \pi/2$, and so the domain of dependence in the limit contains all points in the field with $0 \leqslant t \leqslant T$. Since the solution of the differential equation at the point (X,T) depends on all the data at $t = 0$ together with that portion on the boundary for which $0 \leqslant t \leqslant T$ (initial boundary value problem only), the domain of dependence of the differential

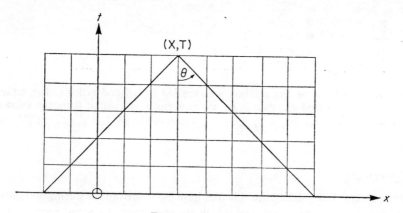

Figure 2

equation consists of the rectangle with dimensions consisting of the initial line in the x- direction and $0 \leqslant t \leqslant T$ in the t- direction. Thus the relationship (41) ensures that the domain of dependence of an explicit scheme like (9) converges to that of the differential equation. In fact, a relationship

$$k = rh^\alpha \quad (\alpha > 1)$$

will ensure the convergence of the domain of dependence of the explicit difference equation. With an implicit scheme, on the other hand, the domain of dependence coincides with that of the differential equation for all h and k.

As an example of a convergence analysis for a difference formula, we turn to the explicit scheme (9). Introduce

$$z_m^n = u_m^n - U_m^n,$$

the difference between the theoretical solutions of the differential and difference equations at the grid point $X = mh$, $T = nk$. From equation (13), this satisfies the equation

$$z_m^{n+1} = (1 - 2r)z_m^n + r(z_{m+1}^n + z_{m-1}^n) + 0(k^2 + kh^2). \tag{42}$$

If

$$0 < r \leqslant \tfrac{1}{2}, \tag{43}$$

the coefficients on the right-hand side of equation (42) are all non-negative and so

$$|z_m^{n+1}| \leqslant (1 - 2r)|z_m^n| + r|z_{m+1}^n| + r|z_{m-1}^n| + A(k^2 + kh^2)$$

$$\leqslant Z^{(n)} + A(k^2 + kh^2),$$

where A depends on the upper bounds for $\partial^2 u / \partial t^2$ and $\partial^4 u / \partial x^4$ (see (13)) and $Z^{(n)}$ is the maximum modulus value of z_m^n over the required range of m. Thus

$$Z^{(n+1)} \leqslant Z^{(n)} + A(k^2 + kh^2),$$

and so if

$$Z^{(0)} = 0$$

(the same initial data for differential and difference equations),

$$\begin{aligned} Z^{(n)} &\leqslant nA(k^2 + kh^2) \\ &= TA(k + h^2) \\ &\to 0 \text{ as } k, h \to 0 \text{ for fixed X, T.} \end{aligned}$$

Convergence has therefore been established if the expression (43) is satisfied.

Exercise

5. Establish the convergence of the backward difference implicit formula (40) for $r > 0$.

Finally, convergence of the Crank–Nicolson implicit scheme (31) can be established for $r > 0$. This is more difficult and the interested reader is referred to page 508 of Isaacson and Keller (1967).

2.7 Stability

We shall first introduce the problem of stability of a finite difference calculation used to solve equation (1) as consisting of finding a condition under which

$$U_m^n - \tilde{U}_m^n \ (\equiv Z_m^n),$$

the difference between the theoretical and numerical solutions of the difference equation, remains bounded as *n tends to infinity*. The latter occurs in a calculation where either

(i) k remains fixed for all n and $t \to \infty$, or
(ii) $h, k \to 0$ $(k/h^2$ fixed) for a fixed value of $t = nk$.

The following methods are used for examining this notion of stability of a finite difference calculation.

The von Neumann method

Here, a harmonic decomposition is made of the error Z at grid points at a given time level, leading to the error function

$$E(x) = \sum_j A_j e^{i\beta_j x},$$

where in general the frequencies $|\beta_j|$ and j are arbitrary. It is necessary to consider only the single term $e^{i\beta x}$ where β is any *real* number. For convenience, suppose that the time level being considered corresponds to $t = 0$. To investigate the error propagation as t increases, it is necessary to find a solution of the finite difference equation which reduces to $e^{i\beta x}$ when $t = 0$. Let such a solution be

$$e^{\alpha t} e^{i\beta x}$$

where $\alpha = \alpha(\beta)$ is, in general, complex. The original error component $e^{i\beta x}$ will not grow with time if

$$|e^{\alpha k}| \leqslant 1$$

for all α. This is von Neumann's criterion for stability.*

---

* In order to allow for exponentially growing solutions of the partial differential equation itself, a more general form is

$$|e^{\alpha k}| \leqslant 1 + 0(k).$$

Example 4   Determine the stability condition for the formula

$$U_m^{n+1} = (1 - 2r)U_m^n + r(U_{m+1}^n + U_{m-1}^n)$$

which is used to solve the equation

$$\frac{\partial u}{\partial t} = \frac{\partial^2 u}{\partial x^2}$$

in region (c) of figure 1.

In this case, let $Mh = 1$. Denote the errors at the grid points in the range $0 \leqslant x \leqslant 1$ at $t = 0$ by $Z(mh) = Z_m^0 (m = 0, 1, \ldots, M)$. Then

$$Z_m^0 = \sum_{j=0}^{M} A_j e^{i\beta_j m h} \quad (m = 0, 1, \ldots, M)$$

determines the unknowns $A_j (j = 0, 1, \ldots, M)$. In this problem, the number of harmonics corresponds to the number of grid points at any time level, and so the number of harmonics increases as the mesh size is reduced. Now, since Z_m^n satisfies the original difference equation, we get

$$Z_m^{n+1} = (1 - 2r)Z_m^n + r(Z_{m+1}^n + Z_{m-1}^n). \tag{44}$$

Substitute

$$Z_m^n = e^{\alpha n k} e^{i\beta m h} = \xi^n e^{i\beta m h}$$

where

$$\xi = e^{\alpha k},$$

and equation (44) gives

$$\xi^{n+1} e^{i\beta m h} = \xi^n [(1 - 2r) e^{i\beta m h} + r(e^{i\beta(m+1)h} + e^{i\beta(m-1)h})].$$

Cancellation of $\xi^n e^{i\beta m h}$ leads to

$$\xi = (1 - 2r) + r(e^{i\beta h} + e^{-i\beta h})$$

$$= 1 - 2r(1 - \cos \beta h)$$

$$= 1 - 4r \sin^2 \frac{\beta h}{2}.$$

The quantity ξ is called the *amplification factor*. The condition for stability, $|\xi| \leqslant 1$, for *all values of* βh, leads to

$$-1 \leqslant 1 - 4r \sin^2 \frac{\beta h}{2} \leqslant +1 \quad (\text{all } \beta h).$$

The right-hand side of the inequality is trivially satisfied if $r > 0$, and the left-hand side gives

40

$$r \leqslant \frac{1}{2 \sin^2 \frac{\beta h}{2}} \; ,$$

leading to the stability condition $0 < r \leqslant \frac{1}{2}$.

The following important points should be noted concerning the von Neumann method of examining stability.

(i) The method which is based on Fourier series applies only if the coefficients of the linear difference equation are constant. If the difference equation has variable coefficients, the method can still be applied locally and it might be expected that a method will be stable if the von Neumann condition, derived as though the coefficients were constant, is satisfied at every point of the field. There is much numerical evidence to support this contention.

(ii) For two level difference schemes with *one dependent variable* and any number of independent variables, the von Neumann condition is sufficient as well as necessary for stability. Otherwise, the condition is *necessary* only.

(iii) Boundary conditions are neglected by the von Neumann method which applies in theory only to pure initial value problems with periodic initial data. It does however provide necessary conditions for stability of constant coefficient problems regardless of the type of boundary condition.

Exercise

6. Find the stability condition for formula (10).

7. Show that the implicit formulae (31) and (32) are unconditionally stable [i.e. stable for all values of $r > 0$].

The matrix method

This is applicable to the initial boundary value problem illustrated in figure 1 (c). If $Mh = 1$, the totality of difference equations connecting values of U at two neighbouring time levels can be written in the matrix form

$$A_n \mathbf{U}^{n+1} = B_n \mathbf{U}^n \tag{45}$$

where the column vector $\mathbf{U}^s$ ($s = n, n + 1$) contains the grid values at the time level s, and A_n, B_n are square matrices of order $(M + 1)$ which may vary with n. If the difference equation is explicit, $A_n = I$ for all n. Also if n tends to infinity while $h, k \to 0$ the order of A_n and B_n tends to infinity.

Equation (45) may be written in the explicit form

$$\mathbf{U}^{n+1} = C_n \mathbf{U}^n$$

where $C_n = A_n^{-1} B_n$, provided $|A_n| \neq 0$. The error vector $\mathbf{Z}^n$ ($\equiv \mathbf{U}^n - \tilde{\mathbf{U}}^n$) therefore satisfies

$$\mathbf{Z}^{n+1} = C_n \mathbf{Z}^n,$$

which leads to

$$\mathbf{Z}^{n+1} = (\prod_{i=0}^{n} C_i)\mathbf{Z}^0,$$

The method defined by (45) will be stable provided the norm $\|\mathbf{Z}^{n+1}\|$ is bounded for all $n \geq 0$. Since

$$\|\mathbf{Z}^{n+1}\| \leq \|\prod_{i=0}^{n} C_i\| \ \|\mathbf{Z}^0\|,$$

this occurs if and only if a constant K (independent of h and k) can be found such that

$$\|\prod_{i=0}^{n} C_i\| \leq K.$$

Now for any matrix C with elements independent of n, the spectral radius (maximum modulus eigenvalue), $\rho(C)$, is related to the norm, $\|C\|$, by the inequality (see Chapter 1)

$$\rho^{n+1}(C) \leq \|C^{n+1}\| \leq \|C\|^{n+1}$$

for all $n \geq 0$. Bearing these inequalities in mind, we consider two simple criteria for regulating the error growth in a calculation based on (45). They are

(i) The spectral radius condition

$$\rho(C) \leq 1, \tag{46a}$$

which is *necessary* for stability and for $\rho(C) < 1$ guarantees that the error vector $\mathbf{Z}^n \to 0$ as $n \to \infty$ (see theorem II, Chapter 1), but gives no indication of the magnitude of $\mathbf{Z}^n$ for finite n, and

(ii) The norm condition

$$\|C\| \leq 1, \tag{46b}$$

which is *sufficient* for stability and guarantees an ever-diminishing error as n increases. (see theorem I, Chapter 1)

As with the von Neumann criterion for stability, if the right hand sides of (46a) and (46b) are replaced by $1 + 0(k)$, exponentially growing solutions of the differential system are allowed. The gap between the conditions (46a) and (46b) depends on the matrix C and we now look at particular types of matrix C.

(i) *Symmetric* Here if we consider the norm $\| \ \|_2$, then

$$\rho(C) = \|C\|_2,$$

and (46a) is necessary and sufficient for stability.

(ii) *Similar to a symmetric matrix* This time, we have

$$P^{-1} CP = \tilde{C}$$

where P is some non-singular matrix, and $\tilde{C}$ is symmetric. Although (46a) is again necessary and sufficient for stability, (proved in example 5) it does not imply (46b) and so the error $\mathbf{Z}^n$ can be large for finite n.

(iii) *Otherwise* If the matrix C is other than (i) or (ii) above we, are unable to comment on the gap between conditions (46a) and (46b).

The reader interested in this material will find numerical experiments in section 6.7, which demonstrate that substantial error growth can arise when C is of type (ii) or (iii) above, even if condition (46a) is satisfied [see also Morton (1979)].

Example 5. Show that a calculation carried out according to (45), with C_n the constant matrix C, is stable if condition (46a) holds and C is similar to a symmetric matrix.

From

$$\tilde{C} = P^{-1} CP \qquad (P \text{ non-singular}),$$

we get

$$\tilde{C}^n = P^{-1} C^n P \qquad (n = 1, 2, \dots),$$

and

$$C^n = P\tilde{C}^n P^{-1}.$$

These results lead to

$$\|\tilde{C}^n\|_2 \leqslant \|P^{-1}\|_2 \ \|P\|_2 \ \|C^n\|_2,$$

and

$$\|C^n\|_2 \leqslant \|P\|_2 \ \|P^{-1}\|_2 \ \|\tilde{C}^n\|_2$$

respectively, and so

$$\frac{\|\tilde{C}^n\|_2}{\|P^{-1}\|_2 \, \|P\|_2} \leqslant \|C^n\|_2 \leqslant \|P\|_2 \, \|P^{-1}\|_2 \, \|\tilde{C}^n\|_2 .$$

But

$$\|\tilde{C}^n\|_2 = \rho(\tilde{C}^n) = \rho^n(\tilde{C}) = \rho^n(C) \leqslant 1,$$

and so provided $\|P\|_2$ and $\|P^{-1}\|_2$ are finite, it follows that $\|C^n\|_2$ is bounded, and so the procedure is stable.

As an example of the matrix method for examining stability, we consider the initial boundary value problem consisting of the heat conduction equation

$$\frac{\partial u}{\partial t} = \frac{\partial^2 u}{\partial x^2},$$

subject to the boundary conditions $u = 0$ at $x = 0,1$ $(t \geqslant 0)$ and the initial condition $u = f(x)$ at $t = 0$ $(0 \leqslant x \leqslant 1)$, being solved by the explicit difference equation

$$U_m^{n+1} = (1 - 2r)U_m^n + r(U_{m+1}^n + U_{m-1}^n).$$

In this case, $A_n = I$ for all n, and

$$B_n = C_n = \begin{bmatrix} (1-2r) & r & & & & \mathbf{O} \\ r & (1-2r) & r & & & \\ & r & \cdot & \cdot & & \\ & & \cdot & \cdot & \cdot & \\ & & & \cdot & \cdot & r \\ \mathbf{O} & & & & r & (1-2r) \end{bmatrix} \quad \text{for all } n.$$

The eigenvalues of this matrix are

$$\lambda_s = 1 - 4r \sin^2 \frac{s\pi}{2M} \, (s = 1, 2, \ldots, M-1),$$

and from equation (46a), it follows that the method is stable if

$$-1 \leqslant 1 - 4r \sin^2 \frac{s\pi}{2M} \leqslant 1 \, (s = 1, 2, \ldots, M-1),$$

which leads to

$$0 < r \leqslant \tfrac{1}{2},$$

an identical result to that obtained by the method of von Neumann.

The energy method

This method of analysing stability may be applied in principle to problems with variable coefficients and to non-linear equations. Each problem requires a different treatment but we can illustrate the general philosophy by means of an example (see Richtmyer and Morton, (1967) p. 13). Consider the problem of solving the heat conduction equation

$$\frac{\partial u}{\partial t} = \frac{\partial^2 u}{\partial x^2} \quad \text{in } (0 \leqslant x \leqslant 1) \times (t \geqslant 0)$$

with homogeneous Dirichlet boundary conditions

$$u(0, t) = u(1, t) = 0 \qquad (t \geqslant 0)$$

and initial data

$$u(x, 0) = f(x) \qquad (0 \leqslant x \leqslant 1).$$

Multiplying the equation by u and integrating with respect to x, we obtain

$$\int_0^1 u \frac{\partial u}{\partial t} dx = \int_0^1 u \frac{\partial^2 u}{\partial x^2} dx,$$

which on integration by parts of the right hand side gives

$$\frac{\partial}{\partial t} \int_0^1 u^2 dx = - \int_0^1 \left(\frac{\partial u}{\partial x} \right)^2 dx \leqslant 0.$$

We deduce from this, that

$$\int_0^1 u^2(x, t)dx \leqslant \int_0^1 u^2(x, 0)dx = \int_0^1 f^2(x)dx$$

and therefore the quantity $\int_0^1 u^2(x, t)dx$ remains bounded as $t \to \infty$.

When the explicit scheme (9) is used, the energy method may be employed to show that the analogous quantity $h \sum_{m+1}^{M-1} (Z_m^n)^2$ remains bounded as $n \to \infty$ as follows:

The nodal errors Z_m^n satisfy the difference equation (see (44))

$$Z_m^{n+1} - Z_m^n = r(Z_{m+1}^n + Z_{m-1}^n - 2Z_m^n) \qquad (m = 1, 2, \ldots, M - 1),$$

and the boundary conditions give

$$Z_0^n = Z_M^n = 0 \qquad \text{(for all } n\text{)}.$$

The difference equation is multiplied by $(Z_m^{n+1} + Z_m^n)$ and the result summed over $m = 1, 2, \ldots, M - 1$ to give

$$\| Z^{n+1} \|^2 - \| Z^n \|^2 = r \sum_{m=1}^{M-1} (Z_m^{n+1} + Z_m^n)(Z_{m+1}^n + Z_{m-1}^n - 2Z_m^n) \qquad (47)$$

where $\| Z^n \|^2 = \sum_{m=1}^{M-1} (Z_m^n)^2$. In order to proceed further we require the identities given in the following exercise.

Exercise

8. Show that

$$\sum_{m=1}^{M-1} V_m (Z_{m-1} - Z_m) = \sum_{m=1}^{M-1} Z_m (V_{m+1} - V_m) \qquad (48a)$$

if $Z_0 = V_M = 0$ and

$$\sum_{m=1}^{M-1} V_m (Z_{m+1} - Z_m) = \sum_{m=1}^{M-1} Z_{m+1} (V_m - V_{m+1}) - V_1 Z_1 \qquad (48b)$$

if $Z_M = 0$.

The right hand side of (47) can be rearranged to read

$$r \sum_{m=1}^{M-1} V_m [(Z_{m+1}^n - Z_m^n) - (Z_m^n - Z_{m-1}^n)]$$

with $V_m = Z_m^{n+1} + Z_m^n$. Substituting (48a) and (48b) into this expression and bearing in mind that

$$Z_0^n = Z_0^{n+1} = Z_M^n = Z_M^{n+1} = 0$$

we find that (47) becomes

$$\| Z^{n+1} \|^2 - \| Z^n \|^2 = -r \left[\sum_{m=0}^{M-1} \left\{ (Z_{m+1}^n - Z_m^n)^2 + \right. \right.$$
$$\left. \left. (Z_{m+1}^{n+1} - Z_m^{n+1})(Z_{m+1}^n - Z_m^n) \right\} \right] \qquad (49)$$

If we define

$$E_n = \| Z^n \|^2 - \tfrac{1}{2} r \sum_{m=0}^{M-1} (Z_{m+1}^n - Z_m^n)^2 \qquad (n = 0, 1, 2, \ldots),$$

it follows that

46

$$E_n \leqslant \| \mathbf{Z}^n \|^2 \quad,$$

and

$$E_{n+1} - E_n = \| \mathbf{Z}^{n+1} \|^2 - \| \mathbf{Z}^n \|^2 - \tfrac{1}{2} r \sum_{m=0}^{M-1} \{ (Z_{m+1}^{n+1} - Z_m^{n+1})^2$$

$$- (Z_{m+1}^n - Z_m^n)^2 \}$$

Substituting for $\| \mathbf{Z}^{n+1} \|^2 - \| \mathbf{Z}^n \|^2$ from (49), we get

$$E_{n+1} - E_n = -\tfrac{1}{2} r \sum_{m=0}^{M-1} (Z_{m+1}^{n+1} - Z_m^{n+1} + Z_{m+1}^n - Z_m^n)^2 \leqslant 0.$$

We conclude therefore that E_n is a monotonic decreasing function of n and it remains to show that $\| \mathbf{Z}^n \|^2$ is bounded as $n \to \infty$.

Summing the inequalities

$$(Z_{m+1}^n - Z_m^n)^2 \leqslant (Z_{m+1}^n - Z_m^n)^2 + (Z_{m+1}^n + Z_m^n)^2 = 2(Z_{m+1}^n)^2 + 2(Z_m^n)^2$$

for $m = 0, 1, \ldots, M-1$, leads to

$$\sum_{m=0}^{M-1} (Z_{m+1}^n - Z_m^n)^2 \leqslant 2 \sum_{m=0}^{M-1} [(Z_{m+1}^n)^2 + (Z_m^n)^2] = 4 \sum_{m=0}^{M-1} (Z_m^n)^2 = 4 \| \mathbf{Z}^n \|^2,$$

and so from the definition of E_n we get

$$E_n \geqslant (1 - 2r) \| \mathbf{Z}^n \|^2.$$

If $0 < r < \tfrac{1}{2}$,

$$\| \mathbf{Z}^n \|^2 \leqslant \frac{1}{1 - 2r} E_n,$$

and in view of the inequalities

$$E_n \leqslant E_{n-1} \leqslant \ldots \leqslant E_0 \leqslant \| \mathbf{Z}^0 \|^2,$$

we have

$$\| \mathbf{Z}^n \|^2 \leqslant \frac{1}{1 - 2r} \| \mathbf{Z}^0 \|^2.$$

The vectors $\mathbf{Z}^n$, $n = 0, 1, 2, \ldots$ are therefore bounded provided $0 < r < \tfrac{1}{2}$ and stability results.

The quantity $\| \mathbf{Z}^n \|^2$ is called the *energy* from which the method gets its name, but it is in no way related to the physical energy of the system. The successful application of this method to more general problems relies heavily on the ingenuity of the user to identify a suitable quantity E_n. For this reason, only *sufficient* conditions for stability can be derived. A comprehensive discussion, along with several examples, may be found in Richtmyer and Morton (1967) (Chapter 6).

Exercise

9. If the Crank–Nicolson formula (31) is used to solve the heat conduction equation, use the method outlined above to show that

$$\| \mathbf{Z}^{n+1} \|^2 - \| \mathbf{Z}^n \|^2 = -r \sum_{m=0}^{M-1} (V_{m+1} - V_m)^2$$

where $V_m = Z_m^{n+1} + Z_m^n$. Deduce directly from this result that the method is stable for $r > 0$.

So far, no mention has been made of the *consistency* of a finite difference approximation to a differential equation. Consistency ensures that the difference equation converges to the correct differential equation as the grid spacing tends to zero. A difference approximation to a parabolic equation is *consistent* if

$$\frac{\text{Truncation error}}{k} \to 0 \text{ as } h, k \to 0.$$

For example, the standard four point explicit formula (9) is consistent since

$$\frac{\text{Truncation error}}{k} = (\tfrac{1}{2}k \frac{\partial^2 u}{\partial t^2} - \tfrac{1}{12}h^2 \frac{\partial^4 u}{\partial x^4} + \dots)$$

$$\to 0 \text{ as } h, k \to 0.$$

In fact, all difference replacements so far mentioned in this book are consistent.

A full discussion of the concepts of stability, convergence and consistency is well beyond the scope of this book and readers who wish to study these topics further are referred to Isaacson and Keller (1967) and Richtmyer and Morton (1967), where there is included a discussion of Lax's equivalence theorem. This theorem states that: 'Given a properly posed linear initial value problem and a finite difference approximation to it that satisfies the consistency condition, stability (as $h, k \to 0$) is the necessary and sufficient condition for convergence'.

Exercise

10. Show that the difference method given by (31) is consistent.

2.8 Derivative boundary conditions

Consider now the equation of heat conduction

$$\frac{\partial u}{\partial t} = \frac{\partial^2 u}{\partial x^2}$$

in the region R, $[0 \leqslant x \leqslant 1] \times [t \geqslant 0]$, subject to the initial condition

$$u(x, 0) = f(x) \qquad (0 \leqslant x \leqslant 1),$$

and the boundary conditions

$$\frac{\partial u}{\partial x} - pu = \lambda_0(t) \qquad (x = 0, t \geqslant 0),$$

and

$$\frac{\partial u}{\partial x} + qu = \lambda_1(t) \qquad (x = 1, t \geqslant 0).$$

The functions $\lambda_0(t)$, $\lambda_1(t)$ are continuous and bounded as $t \to \infty$. This system constitutes the *third* initial boundary value problem for the heat conduction equation. The particular case $p = q = 0$ gives the Neumann, or second, initial boundary value problem. The Dirichlet, or first, problem with u prescribed on the boundaries $x = 0, 1$ ($t \geqslant 0$) has been considered earlier in the chapter.

The finite difference formula used to solve this problem is the Crank–Nicolson formula (31), which is stable for all $r > 0$. The derivatives in the boundary conditions are replaced by

$$\left(\frac{\partial u}{\partial x}\right)_m^s = \frac{1}{2h}(U_{m+1}^s - U_{m-1}^s) \quad (m = 0, M; s = n, n+1), \tag{50}$$

where $Mh = 1$. At grid points on the boundaries, equations (31) and (50) are used to eliminate U_{-1}^s and U_{M+1}^s respectively where $s = n, n+1$, and so the totality of equations can be expressed in the form

$$A\mathbf{U}^{n+1} = B\mathbf{U}^n + \mathbf{k}^n \tag{51}$$

where $\mathbf{U}^s$ ($s = n, n+1$) denotes the column vector

$$[U_0^s, U_1^s, \ldots, U_M^s]^T,$$

$$A = I - \tfrac{1}{2}rQ, \qquad B = I + \tfrac{1}{2}rQ,$$

where Q is a matrix of order $(M + 1)$ given by

$$Q = \begin{bmatrix} -2\left(1+\dfrac{p}{M}\right) & 2 & & & & & \bigcirc \\ 1 & -2 & 1 & & & & \\ & & \ddots & \ddots & \ddots & & \\ & & & 1 & -2 & 1 \\ \bigcirc & & & & 2 & -2\left(1+\dfrac{q}{M}\right) \end{bmatrix}$$

and I is the unit matrix of order $M + 1$. The vector $\mathbf{k}^n$ has $(M + 1)$ components involving the boundary conditions.

In order to study the stability of equation (51), it is written in the form

$$\mathbf{U}^{n+1} = A^{-1}B\mathbf{U}^n + A^{-1}\mathbf{k}^n,$$

where it is assumed that $|A| \neq 0$. In fact,

$$A^{-1}B = (I - \tfrac{1}{2}rQ)^{-1}(I + \tfrac{1}{2}rQ),$$

and is not symmetric. Introduce the diagonal matrix

$$D = \begin{bmatrix} \sqrt{2} & & & & \\ & 1 & & \bigcirc & \\ & & \ddots & & \\ & \bigcirc & & 1 & \\ & & & & \sqrt{2} \end{bmatrix}$$

of order $(M + 1)$, so that Q is similar to the symmetric matrix

$$\widetilde{Q} = D^{-1}QD.$$

Then

$$\begin{aligned}
(\widetilde{A^{-1}B}) &= D^{-1}(A^{-1}B)D \\
&= D^{-1}(I - \tfrac{1}{2}rQ)^{-1}(I + \tfrac{1}{2}rQ)D \\
&= [D^{-1}(I - \tfrac{1}{2}rQ)^{-1}D][D^{-1}(I + \tfrac{1}{2}rQ)D] \\
&= [D^{-1}(I - \tfrac{1}{2}rQ)D]^{-1}[D^{-1}(I + \tfrac{1}{2}rQ)D] \\
&= [I - \tfrac{1}{2}r\widetilde{Q}]^{-1}[I + \tfrac{1}{2}r\widetilde{Q}].
\end{aligned}$$

But the matrices $[I - \tfrac{1}{2}r\widetilde{Q}]^{-1}$ and $[I + \tfrac{1}{2}r\widetilde{Q}]$ are symmetric and commute, and so $(\widetilde{A^{-1}B})$ is symmetric. Thus $(A^{-1}B)$ is similar to the symmetric matrix $(\widetilde{A^{-1}B})$

and so

$$\rho(A^{-1}B) \equiv \rho(\widehat{A^{-1}B}) \leqslant 1 \tag{52}$$

is a necessary and sufficient condition for stability, where $\rho(A^{-1}B)$ is the spectral radius of $A^{-1}B$. (see example (5)).

Now the eigenvalues μ_j $(j = 0, 1, \ldots, M)$ of $A^{-1}B$ are given by

$$\mu_j = (1 - \tfrac{1}{2}r\lambda_j)^{-1}(1 + \tfrac{1}{2}r\lambda_j), \tag{53}$$

where λ_j $(j = 0, 1, \ldots, M)$ are the eigenvalues of the matrix Q. Since

$$\rho(A^{-1}B) = \max_j |\mu_j|,$$

the condition for stability (52), together with (53), gives

$$\lambda_j \leqslant 0 \text{ (all } j). \tag{54}$$

The eigenvalues $\lambda_j(j = 0, 1, \ldots, M)$ of Q are obtained from the equation

$$|Q - \lambda I| = 0.$$

Expanding the determinant by the elements of the first and last rows respectively, after a certain amount of manipulation, the result

$$f(\lambda) \equiv \left[(\lambda + 2)^2 + \frac{2(p + q)}{M}(\lambda + 2) - 4\left(1 - \frac{pq}{M^2}\right) \right] T_{M-1}(\lambda) +$$

$$4 \frac{p + q}{M} T_{M-2}(\lambda) = 0 \tag{55}$$

is obtained, where $M \geqslant 3$, and

$$T_M(\lambda) = \begin{vmatrix} -(2 + \lambda) & 1 & & & & \\ 1 & -(2 + \lambda) & 1 & & \mathbf{O} & \\ \cdot & \cdot & \cdot & & & \\ & \cdot & \cdot & \cdot & & \\ \mathbf{O} & & \cdot & & \cdot & 1 \\ & & & & 1 & -(2 + \lambda) \end{vmatrix}$$

$$= (-1)^M \prod_{j+1}^{M}\left(4 \cos^2 \frac{\pi j}{2(M + 1)} + \lambda\right). \tag{56}$$

Two special cases are now considered

(i) $p = q = \infty$. This is equivalent to $u = 0$ on $x = 0, 1$ which constitutes Dirichlet boundary conditions. In this case, the result (55) reduces to

$$T_{M-1}(\lambda) = 0,$$

which, from (56), gives

$$\prod_{j=1}^{M-1} \left(4 \cos^2 \frac{\pi j}{2M} + \lambda \right) = 0,$$

and so

$$\lambda_j = -4 \cos^2 \frac{\pi j}{2M} \ (j = 1, 2, \ldots, M-1).$$

From (54), it follows that this case is stable. It should be noted that this system has $(M - 1)$ eigenvalues μ_j, which is a reduction of two compared with the general case.

(ii) $p = q = 0$ (Neumann problem). This time, equation (55) reduces to

$$[(\lambda + 2)^2 - 4] T_{M-1}(\lambda) = 0,$$

and so

$$\lambda = -4 \cos^2 \frac{\pi j}{2M} \ (j = 0, 1, 2, \ldots, M-1, M)$$

This case is also stable, although it has one root zero. In fact, the problem with

$$p = q = 0,$$

$$\lambda_0(t) = \pi e^{-\pi^2 t},$$

$$\lambda_1(t) = -\pi e^{-\pi^2 t},$$

$$f(x) = \sin \pi x \quad (0 \leqslant x \leqslant 1)$$

was solved numerically using the Crank–Nicolson method with $r = \frac{1}{2}$ and $M = 10$. The theoretical solution of the differential system is

$$u = e^{-\pi^2 t} \sin \pi x,$$

and the errors (E) at $x = 0.5$ are shown in table 2 after a number of time steps (P). A *persistent error* is detected from the numerical results and this is due to the presence of a root $\lambda = 0$.

Table 2

P	E
20	− 0.000,828
40	− 0.003,616
60	− 0.004,718
80	− 0.005,130
100	− 0.005,284
120	− 0.005,341
140	− 0.005,362
160	− 0.005,370
180	− 0.005,373
200	− 0.005,374

For $r > \frac{1}{2}$, we see from (53) that some or all of the eigenvalues of $A^{-1}B$ may be negative and consequently the numerical solution based on (51) will contain oscillations even though it is stable. This Crank–Nicolson 'noise' is caused by the larger values of $|\lambda_j|$ and is usually damped out after a few time steps. A further discussion of Crank–Nicolson noise can be found in Wood and Lewis (1975).

Returning to the case of general p, q it is interesting to determine the conditions on p, q for a problem to have a root $\lambda = 0$. From equation (55), we see that

$$f(0) \equiv 4\left[\left\{\frac{p+q}{M} + \frac{pq}{M^2}\right\} T_{M-1}(0) + \frac{p+q}{M} T_{M-2}(0)\right] = 0,$$

where, from equation (56)

$$T_M(0) = (-1)^M \prod_{j=1}^{M}\left(4\cos^2 \frac{\pi j}{2(M+1)}\right) = (-1)^M(M+1)$$

and so

$$f(0) = (-1)^{M-1}\, 4\left(\frac{p+q+pq}{M}\right) = 0,$$

leading to the condition

$$p + q + pq = 0, \tag{57}$$

for a problem to have a root $\lambda = 0$. The Neumann problem ($p = q = 0$) is a special case of condition (57). Any problem with p, q satisfying condition (57) will have a persistent error.

To facilitate examination of equation (55), it is convenient to write

$$T_M(\lambda) = \frac{\sin (M+1)\phi}{\sin \phi}, \tag{58}$$

where $\cos \phi = -\frac{1}{2}(2 + \lambda)$. Using equations (55) and (58) the following additional properties can be obtained in the general case.

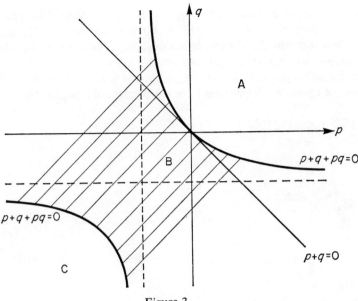

Figure 3

(i) For $p + q + pq > 0$ and $p + q > 0$ (region A of figure 3), all roots are negative.

(ii) For $p + q + pq > 0$ and $p + q < 0$ (region C), there are two positive roots.

(iii) For $p + q + pq < 0$ (region B), there is one positive root.

These positive roots correspond to exponentially growing solutions in the differential equation.

A more detailed investigation of derivative boundary conditions can be found in Keast and Mitchell, (1967).

2.9 Parabolic equations in two space dimensions

We now consider finite difference methods of solution of the equation

$$\frac{\partial u}{\partial t} = Lu, \tag{59}$$

where

$$L \equiv \sum_{i=1}^{2} \left[\frac{\partial}{\partial x_i} \left(a_i(x_1, x_2, t) \frac{\partial}{\partial x_i} \right) + b_i(x_1, x_2, t) \frac{\partial}{\partial x_i} \right] - c(x_1, x_2, t) \tag{60}$$

with

(i) a_1, a_2 strictly positive, and

(ii) c non-negative.

C

Again the region to be examined in (x_1, x_2, t) space is covered by a rectilinear grid with sides parallel to the axes, with h, k the grid spacings in the space and time directions respectively. The grid points (X_1, X_2, T) are given by $X_1 = lh$, $X_2 = mh$, $T = nk$, where l, m, n are integers and $l = m = n = 0$ is the origin. The function satisfying the difference equation at the grid points is $U_{l,m}^n$. The exact difference replacement of equation (59), provided L is independent of t, is given by

$$U_{l,m}^{n+1} = \exp (kL) U_{l,m}^n, \qquad (61)$$

with

$$\frac{\partial}{\partial x_1} = \frac{2}{h} \sinh^{-1} \frac{\delta_{x_1}}{2}$$

and

$$\frac{\partial}{\partial x_2} = \frac{2}{h} \sinh^{-1} \frac{\delta_{x_2}}{2},$$

where

$$\delta_{x_1} U_{l,m}^n = U_{l+1/2,m}^n - U_{l-1/2,m}^n$$

and

$$\delta_{x_2} U_{l,m}^n = U_{l,m+1/2}^n - U_{l,m-1/2}^n.$$

In the following sections, U^n will be written in short for $U_{l,m}^n$

2.10 Explicit methods

It is advantageous to examine the simplest possible case of L, which arises from the standard heat conduction equation, viz.

$$L \equiv \frac{\partial^2}{\partial x_1^2} + \frac{\partial^2}{\partial x_2^2} \equiv D_1^2 + D_2^2,$$

where

$$D_1 \equiv \frac{\partial}{\partial x_1} \quad \text{and} \quad D_2 \equiv \frac{\partial}{\partial x_2}.$$

Here equation (61) becomes

$$U^{n+1} = \exp (kD_1^2) \exp (kD_2^2) U^n,$$

provided D_1^2 and D_2^2 commute, where

$$D_1^2 = \frac{1}{h^2} (\delta_{x_1}^2 - \tfrac{1}{12}\delta_{x_1}^4 + \tfrac{1}{90}\delta_{x_1}^6 \dots)$$

and

$$D_2^2 = \frac{1}{h^2} (\delta_{x_2}^2 - \tfrac{1}{12}\delta_{x_2}^4 + \tfrac{1}{90}\delta_{x_2}^6 \dots).$$

Elimination of D_1^2 and D_2^2 leads to

$$U^{n+1} = [1 + r\delta_{x_1}^2 + \tfrac{1}{2}r(r - \tfrac{1}{6})\delta_{x_1}^4 \dots] [1 + r\delta_{x_2}^2 + \tfrac{1}{2}r(r - \tfrac{1}{6})\delta_{x_2}^4 \dots] U^n, \quad (62)$$

where $r = k/h^2$. Various explicit difference formulae can be obtained from equation (62). For example,

$$U^{n+1} = [1 + r(\delta_{x_1}^2 + \delta_{x_2}^2)] U^n + 0(k^2 + kh^2) \tag{63}$$

is the standard explicit replacement, involving five points at the time level $t = nk$. Another simple formula obtained from equation (62) is

$$U^{n+1} = (1 + r\delta_{x_1}^2)(1 + r\delta_{x_2}^2)U^n + 0(k^2 + kh^2). \tag{64}$$

This formula involves nine points at $t = nk$.

Exercise

11. Obtain the truncation errors of formulae (63) and (64) and show that when $r = \tfrac{1}{6}$, formula (64) has a truncation error

$$0(k^3 + kh^4).$$

The high accuracy formula

$$U^{n+1} = (1 + \tfrac{1}{6}\delta_{x_1}^2)(1 + \tfrac{1}{6}\delta_{x_2}^2)U^n + 0(k^3 + kh^4) \tag{65}$$

is of limited use since the step forward in time is given by $k = \tfrac{1}{6}h^2$, and so if $h = 0.1$, it would require six hundred time steps to advance one unit in time. However, if a solution were required at small values of the time, perhaps to provide extra starting values for a difference method involving three or more time levels, formula (65) might prove useful.

No mention has been made so far of the region to be examined in (x_1, x_2, t) space. Explicit formulae are essential in pure initial value problems, where initial data is given for $-\infty < x_1, x_2 < +\infty$. If the problem is an initial boundary value problem,

56

the region consists of $R \times [t \geq 0]$ where R is an arbitrary closed region in x_1, x_2 space. When R is rectangular, with sides parallel to the x_1- and x_2-axes, formulae such as (63) and (64) can be used at all internal grid points of R. When R is non-rectangular, internal grid points adjacent to the boundary require special treatment. For example, the difference equation valid at a grid point such as P in figure 4 is not given by formula (63), but by

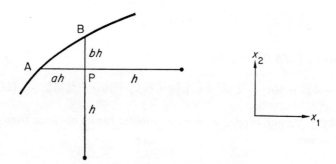

Figure 4

$$U_{l,m}^{n+1} = \left\{1 - 2r\left(\frac{1}{a} + \frac{1}{b}\right)\right\}U_{l,m}^{n} + 2r\left[\frac{1}{1+a}\,U_{l+1,m}^{n} + \frac{1}{a(1+a)}\,U_{l-a,m}^{n} + \right.$$

$$\left. \frac{1}{b(1+b)}\,U_{l,m+b}^{n} + \frac{1}{1+b}\,U_{l,m-1}^{n}\right] + 0(k^2 + kh)$$

where

$$AP = ah, \qquad BP = bh \quad (0 < a, b < 1).$$

Exercise

12. With reference to figure 4, show, using Taylor expansions, that

$$\frac{2}{1+a}\,U_{l+1,m}^{n} + \frac{2}{a(1+a)}\,U_{l-a,m}^{n} - \frac{2}{a}U_{l,m}^{n} = h^2\left(\frac{\partial^2 u}{\partial x^2}\right)_{l,m}^{n} + 0(h^3)$$

and

$$\frac{2}{b(1+b)}\,U_{l,m+b}^{n} + \frac{2}{1+b}\,U_{l,m-1}^{n} - \frac{2}{b}\,U_{l,m}^{n} = h^2\left(\frac{\partial^2 u}{\partial y^2}\right)_{l,m}^{n} + 0(h^3),$$

where $0 < a, b < 1$.

Returning to equation (59), another equation of importance is

$$\frac{\partial u}{\partial t} = \frac{\partial}{\partial x_1}\left(a_1(x_1, x_2, t)\frac{\partial u}{\partial x_1}\right) + \frac{\partial}{\partial x_2}\left(a_2(x_1, x_2, t)\frac{\partial u}{\partial x_2}\right).$$

Here

$$L \equiv D_1(a_1 D_1) + D_2(a_2 D_2),$$

and so

$$U^{n+1} = \exp\left(kD_1(a_1 D_1)\right)\exp\left(kD_2(a_2 D_2)\right)U^n. \tag{66}$$

If we use the approximations

$$D_1(a_1 D_1) = \frac{1}{h^2}\delta_{x_1}(a_1 \delta_{x_1})$$

and

$$D_2(a_2 D_2) = \frac{1}{h^2}\delta_{x_2}(a_2 \delta_{x_2}),$$

where

$$\delta_{x_1}(a_1 \delta_{x_1})U_{l,m}^n = (a_1)_{l+1/2,m}^n(U_{l+1,m}^n - U_{l,m}^n) - (a_1)_{l-1/2,m}^n(U_{l,m}^n - U_{l-1,m}^n)$$

and

$$\delta_{x_2}(a_2 \delta_{x_2})U_{l,m}^n = (a_2)_{l,m+1/2}^n(U_{l,m+1}^n - U_{l,m}^n) - (a_2)_{l,m-1/2}^n(U_{l,m}^n - U_{l,m-1}^n),$$

and expand the right-hand side of equation (66), the formula

$$U_{l,m}^{n+1} = [1 - r\{(a_1)_{l+1/2,m}^n + (a_1)_{l-1/2,m}^n + (a_2)_{l,m+1/2}^n + (a_2)_{l,m-1/2}^n\}]U_{l,m}^n +$$

$$r(a_1)_{l+1/2,m}^n U_{l+1,m}^n + r(a_1)_{l-1/2,m}^n U_{l-1,m}^n + r(a_2)_{l,m+1/2}^n U_{l,m+1}^n +$$

$$r(a_2)_{l,m-1/2}^n U_{l,m-1}^n + O(k^2 + kh^2) \tag{67}$$

can be obtained, which involves five grid points at the time level $t = nk$. This, of course, reduces to formula (63) if $a_1 = a_2 = 1$. A formula involving nine grid points at $t = nk$ is obtained from

$$U^{n+1} = [1 + r\delta_{x_1}(a_1 \delta_{x_1})][1 + r\delta_{x_2}(a_2 \delta_{x_2})]U^n.$$

It consists of formula (60) with the additional term

$$r^2[(a_1 a_2)_{l+1/2,m+1/2}^n(U_{l+1,m+1}^n + U_{l,m}^n - U_{l,m+1}^n - U_{l+1,m}^n) +$$

58

$$(a_1 a_2)_{l-1/2, m-1/2}^{n}(U_{l,m}^{n} + U_{l-1,m-1}^{n} - U_{l-1,m}^{n} - U_{l,m-1}^{n}) +$$

$$(a_1 a_2)_{l-1/2, m+1/2}^{n}(U_{l,m+1}^{n} + U_{l-1,m}^{n} - U_{l-1,m+1}^{n} - U_{l,m}^{n}) -$$

$$(a_1 a_2)_{l+1/2, m-1/2}^{n}(U_{l+1,m}^{n} + U_{l,m-1}^{n} - U_{l,m}^{n} - U_{l+1,m-1}^{n})] \qquad (68)$$

on the right-hand side. This reduces to (63) if $a_1 = a_2 = 1$.

2.11 Stability of explicit methods

The von Neumann method requires a harmonic decomposition of the errors at a given time level, say $t = 0$. The growth of a single term in the double sum representing the decomposition is given by

$$e^{\alpha t} e^{i\beta x_1} e^{i\gamma x_2}, \qquad (69)$$

where β, γ are arbitrary real numbers and $\alpha \equiv \alpha(\beta, \gamma)$ is in general complex. The original error component $e^{i\beta x_1} e^{i\gamma x_2}$ will not grow with time if

$$|e^{\alpha k}| \leqslant 1$$

for all α.

Example 6. Show that formula (63) is stable for $r \leqslant \frac{1}{4}$.

If (69) is substituted into formula (63) and the result divided by

$$e^{\alpha n k} e^{i\beta l h} e^{i\gamma m h},$$

this leads to

$$\xi = e^{\alpha k} = 1 - 4r \left(\sin^2 \frac{\beta h}{2} + \sin^2 \frac{\gamma h}{2} \right).$$

For stability $|\xi| \leqslant 1$ and so

$$-1 \leqslant 1 - 4r \left(\sin^2 \frac{\beta h}{2} + \sin^2 \frac{\gamma h}{2} \right) \leqslant 1.$$

The right-hand side of the inequality is trivially satisfied, and the left-hand side gives

$$r \leqslant \frac{1}{2 \left(\sin^2 \dfrac{\beta h}{2} + \sin^2 \dfrac{\gamma h}{2} \right)}.$$

Since β and γ are completely arbitrary, this leads to the stability condition $r \leqslant \frac{1}{4}$.

Von Neumann's method applied in a similar manner shows that formula (63) has a stability range $0 < r \leqslant \frac{1}{2}$. This method of *improving stability by the addition of a perturbing term*, in this case $r^2 \delta_{x_1}^2 \delta_{x_2}^2 U_{lm}^n$, which is of a higher order than the original formula, is well known, and is used repeatedly in improving the stability of difference approximations.

Exercise

13. Show that formula (63) is stable if

$$0 < r \leqslant \frac{1}{2}.$$

The stability of schemes (67), and (67) modified by the term (68), is not easy to establish for arbitrary functions $a_1(x_1, x_2, t)$ and $a_2(x_1, x_2, t)$. Von Neumann's method, which depends on constant coefficients, does not strictly apply, whereas the matrix method, based on the matrix equation (45) with C_n neither symmetric nor similar to a symmetric matrix, is beyond the scope of this book. An heuristic method based on using von Neumann's method locally at different parts of the region under consideration is often applied successfully.

2.12 Alternating direction implicit (A.D.I.) methods

Because of poor stability properties, explicit difference methods are rarely used to solve initial boundary value problems in two or more space dimensions. Implicit methods with their superior stability properties are almost always used. Unfortunately, an implicit method in two space dimensions requires a set of equations to be solved at the advanced time level, which is not always easy to accomplish directly. Accordingly, alternating direction implicit (A.D.I.) methods are introduced which are two-step methods involving the solution of tridiagonal sets of equations along lines parallel to the x_1- and x_2-axes at the first and second steps respectively.

A.D.I. methods will be illustrated first with respect to equation (59) with

$$L \equiv D_1^2 + D_2^2.$$

The region under consideration in (x_1, x_2, t) space consists of $R \times [t \geqslant 0]$ where R is an arbitrary closed region in x_1, x_2 space and ∂R is its boundary. Unless otherwise stated, *we shall assume that u is given on the boundary of the open region $R \times [t \geqslant 0]$*. *Initially R is the square* $[0 \leqslant x_1, x_2 \leqslant 1]$. The exact difference replacement formula (61) can be written in the form

$$\exp\left[-\frac{1}{2}k(D_1^2 + D_2^2)\right]U^{n+1} = \exp\left[\frac{1}{2}k(D_1^2 + D_2^2)\right]U^n \qquad (70)$$

where

$$D_1^2 = \frac{1}{h^2}(\delta_{x_1}^2 - \tfrac{1}{12}\delta_{x_1}^4 + \tfrac{1}{90}\delta_{x_1}^6 \ldots)$$

and

$$D_2^2 = \frac{1}{h^2}(\delta_{x_2}^2 - \tfrac{1}{12}\delta_{x_2}^4 + \tfrac{1}{90}\delta_{x_2}^6 \ldots).$$

If difference formulae, correct to second differences, are substituted for D_1^2 and D_2^2, equation (70) can be rewritten as

$$\exp\left(-\tfrac{1}{2}r\delta_{x_1}^2\right)\exp\left(-\tfrac{1}{2}r\delta_{x_2}^2\right)U^{n+1} = \exp\left(\tfrac{1}{2}r\delta_{x_1}^2\right)\exp\left(\tfrac{1}{2}r\delta_{x_2}^2\right)U^n,$$

and on expansion, the formula

$$(1 - \tfrac{1}{2}r\delta_{x_1}^2)(1 - \tfrac{1}{2}r\delta_{x_2}^2)U^{n+1} = (1 + \tfrac{1}{2}r\delta_{x_1}^2)(1 + \tfrac{1}{2}r\delta_{x_2}^2)U^n + 0(k^3 + kh^2) \quad (71)$$

is obtained. This is the analogue of the Crank–Nicolson formula (31) in two space dimensions. If an intermediate value $U^{n+1}*$ is introduced, formula (71) can be *split* into the two formulae,

and
$$\left.\begin{array}{l}(1 - \tfrac{1}{2}r\delta_{x_1}^2)U^{n+1}* = (1 + \tfrac{1}{2}r\delta_{x_2}^2)U^n \\[2mm] (1 - \tfrac{1}{2}r\delta_{x_2}^2)U^{n+1} = (1 + \tfrac{1}{2}r\delta_{x_1}^2)U^{n+1}*.\end{array}\right\} \qquad (71a)$$

The split formula (71a) with $U^{n+1}* = U^{n+1/2}$ was first introduced by Peaceman and Rachford (1955) and is known as the *Peaceman–Rachford formula*.

A higher accuracy split formula can be obtained from formula (70) by substituting

$$D_1^2 = \frac{1}{h^2}\delta_{x_1}^2(1 + \tfrac{1}{12}\delta_{x_1}^2)^{-1}$$

and

$$D_2^2 = \frac{1}{h^2}\delta_{x_2}^2(1 + \tfrac{1}{12}\delta_{x_2}^2)^{-1}$$

and by expanding to give

$$[1 - \tfrac{1}{2}(r - \tfrac{1}{6})\delta_{x_1}^2]\,[1 - \tfrac{1}{2}(r - \tfrac{1}{6})\delta_{x_2}^2]\,U^{n+1} =$$

$$[1 + \tfrac{1}{2}(r + \tfrac{1}{6})\delta_{x_1}^2]\,[1 + \tfrac{1}{2}(r + \tfrac{1}{6})\delta_{x_2}^2]\,U^n + 0(k^3 + kh^4), \qquad (72)$$

which can be split into the two formulae

$$[1 - \tfrac{1}{2}(r - \tfrac{1}{6})\delta_{x_1}^2] U^{n+1} * = [1 + \tfrac{1}{2}(r + \tfrac{1}{6})\delta_{x_2}^2] U^n$$

and

$$[1 - \tfrac{1}{2}(r - \tfrac{1}{6})\delta_{x_2}^2] U^{n+1} = [1 + \tfrac{1}{2}(r + \tfrac{1}{6})\delta_{x_1}^2] U^{n+1} *. \qquad (72a)$$

The split formula (72a) was first obtained by Mitchell and Fairweather (1964). Formula (72) is an extension of the Douglas formula (32) in two space dimensions. Since the split formulae (71a) and (72a) both represent two-step methods involving the solution of tridiagonal sets of equations along lines parallel to the x_1- and x_2-axes at the first and second steps respectively, they qualify as A.D.I. methods.

Exercise

14. Show that if

$$U^{n+1} * = U^{n+1/2},$$

the individual formulae in (71a) are consistent, but those in (72a) are *not* consistent.

Formulae (71) and (72) can be split in an alternative manner suggested by D'Yakonov (1963) to give

$$(1 - \tfrac{1}{2}r\delta_{x_1}^2) U^{n+1} * = (1 + \tfrac{1}{2}r\delta_{x_1}^2)(1 + \tfrac{1}{2}r\delta_{x_2}^2) U^n$$

and

$$(1 - \tfrac{1}{2}r\delta_{x_2}^2) U^{n+1} = U^{n+1} *, \qquad (71b)$$

and

$$[1 - \tfrac{1}{2}(r - \tfrac{1}{6})\delta_{x_1}^2] U^{n+1} * = [1 + \tfrac{1}{2}(r + \tfrac{1}{6})\delta_{x_1}^2] [1 + \tfrac{1}{2}(r + \tfrac{1}{6})\delta_{x_2}^2] U^n$$

and

$$[1 - \tfrac{1}{2}(r - \tfrac{1}{6})\delta_{x_2}^2] U^{n+1} = U^{n+1} * \qquad (72b)$$

respectively.

Finally, Douglas and Rachford (1956) formulated an A.D.I. method which is given in split form by

$$(1 - r\delta_{x_1}^2) U^{n+1} * = (1 + r\delta_{x_2}^2) U^n$$

and

$$(1 - r\delta_{x_2}^2) U^{n+1} = U^{n+1} * - r\delta_{x_2}^2 U^n, \qquad (73a)$$

and is known as the *Douglas–Rachford* method. Elimination of $U^{n+1} *$ leads to the formula

$$(1 - r\delta_{x_1}^2)(1 - r\delta_{x_2}^2) U^{n+1} = (1 + r^2\delta_{x_1}^2 \delta_{x_2}^2) U^n, \qquad (73)$$

which can be split in an alternative manner according to the method of D'Yakonov to give

$$(1 - r\delta_{x_1}^2)U^{n+1*} = (1 + r^2\delta_{x_1}^2\delta_{x_2}^2)U^n$$

and

$$(1 - r\delta_{x_2}^2)U^{n+1} = U^{n+1*}.$$

$\left.\begin{array}{c}\\\\\end{array}\right\}$ (73b)

Exercise

15. Show that formula (73) has a truncation error

$$0(k^2 + kh^2)$$

and that if

$$U^{n+1*} = U^{n+1},$$

the individual formulae in (73a) are consistent.

The intermediate value U^{n+1*} introduced in each A.D.I. method is not necessarily an approximation to the solution at any value of the time. As a result, particularly with the high accuracy methods, the boundary values at the intermediate level must be obtained, if possible, in terms of the boundary values at $t = nk$ and $t = (n + 1)k$. The following list of formulae gives U^{n+1*} explicitly in terms of central differences of $g_{l,m}^{n+1}$ and $g_{l,m}^n$ with respect to x_2 only, where

$$u(x_1, x_2, t) = g(x_1, x_2, t)$$

when (x_1, x_2, t) is on the boundary $\partial R \times [t \geq 0]$. The number before the formula refers to the number of the corresponding split formula and the letters P.R., D.R. and D. refer to Peaceman–Rachford, Douglas–Rachford and D'Yakonov type splittings respectively.

(P.R.) $\quad U^{n+1*} = \frac{1}{2}(1 - \frac{1}{2}r\delta_{x_2}^2)g_{l,m}^{n+1} + \frac{1}{2}(1 + \frac{1}{2}r\delta_{x_2}^2)g_{l,m}^n$ (71a)

(D.) $\quad U^{n+1*} = (1 - \frac{1}{2}r\delta_{x_2}^2)g_{l,m}^{n+1}$ (71b)

(P.R.) $\quad U^{n+1*} = \frac{r - \frac{1}{6}}{2r}[1 - \frac{1}{2}(r - \frac{1}{6})\delta_{x_2}^2]g_{l,m}^{n+1} + \frac{r + \frac{1}{6}}{2r}[1 + \frac{1}{2}(r + \frac{1}{6})\delta_{x_2}^2]g_{l,m}^n$

(D.) $\quad U^{n+1*} = [1 - \frac{1}{2}(r - \frac{1}{6})\delta_{x_2}^2]g_{l,m}^{n+1}$ (72b)

(D.R.) $\quad U^{n+1*} = (1 - r\delta_{x_2}^2)g_{l,m}^{n+1} + r\delta_{x_2}^2 g_{l,m}^n$ (73a)

(D.) $\quad U^{n+1*} = (1 - r\delta_{x_2}^2)g_{l,m}^{n+1}.$ (73b)

The D'Yakonov and Douglas–Rachford type splittings already give U^{n+1*} explicitly, whereas the Peaceman–Rachford type splitting requires a small amount of manipulation in order to give U^{n+1*} in terms of $g_{l,m}^{n+1}$ and $g_{l,m}^n$. This is left as an exercise for the reader. The intermediate boundary values given by formulae (71a) and (73a) compare with $g_{l,m}^{n+1/2}$ and $g_{l,m}^{n+1}$, the values given originally for the P.R. and D.R. methods.

If the boundary conditions are *independent of the time*, the formulae giving $U^{n+1}{}^*$ on the boundary reduce to

$$U^{n+1}{}^* = g_{l,m} \qquad \text{(P.R.)}$$

$$U^{n+1}{}^* = (1 - \tfrac{1}{2}r\delta_{x_2}^2)g_{l,m} \qquad \text{(D.)}$$

$$U^{n+1}{}^* = (1 + \tfrac{1}{6}\delta_{x_2}^2)g_{l,m} \qquad \text{(P.R.)}$$

$$U^{n+1}{}^* = [1 - \tfrac{1}{2}(r - \tfrac{1}{6})\delta_{x_2}^2]g_{l,m} \qquad \text{(D.)}$$

$$U^{n+1}{}^* = g_{l,m} \qquad \text{(D.R.)}$$

$$U^{n+1}{}^* = (1 - r\delta_{x_2}^2)g_{l,m} \qquad \text{(D.)}$$

respectively. A more detailed investigation of intermediate boundary values in A.D.I. methods is given in Fairweather and Mitchell (1967).

The stability of the A.D.I. methods of this section for all values of $r > 0$ is easily demonstrated using von Neumann's method. On elimination of $U^{n+1}{}^*$, the methods reduce to one or other of formulae (71), (72), and (73). These are all special cases of the general formula

$$(1 - a\delta_{x_1}^2)(1 - a\delta_{x_2}^2)U^{n+1} = [1 + b(\delta_{x_1}^2 + \delta_{x_2}^2) + c\delta_{x_1}^2\delta_{x_2}^2]U^n, \qquad (74)$$

with a, b and c taking particular values in each case. The corresponding error equation is

$$(1 - a\delta_{x_1}^2)(1 - a\delta_{x_2}^2)Z_{l,m}^{n+1} = [1 + b(\delta_{x_1}^2 + \delta_{x_2}^2) + c\delta_{x_1}^2\delta_{x_2}^2]Z_{l,m}^n. \qquad (75)$$

Substitute

$$Z_{l,m}^n = e^{\alpha nk}\, e^{i\beta lh}\, e^{i\gamma mh},$$

and equation (75) gives the amplification factor

$$e^{\alpha k} = \frac{1 - b(S_\beta^2 + S_\gamma^2) + cS_\beta^2 S_\gamma^2}{(1 + aS_\beta^2)(1 + aS_\gamma^2)}$$

where

$$S_\beta^2 = 4\sin^2\frac{\beta x_1}{2}, \qquad S_\gamma^2 = 4\sin^2\frac{\gamma x_2}{2}.$$

For stability

$$-1 \leqslant \frac{1 - b(S_\beta^2 + S_\gamma^2) + cS_\beta^2 S_\gamma^2}{(1 + aS_\beta^2)(1 + aS_\gamma^2)} \leqslant +1,$$

which leads to

$$-(a - b)(S_\beta^2 + S_\gamma^2) - (a^2 + c)S_\beta^2 S_\gamma^2 \leqslant 2 \tag{76a}$$

and

$$-(a + b)(S_\beta^2 + S_\gamma^2) - (a^2 - c)S_\beta^2 S_\gamma^2 \leqslant 0. \tag{76b}$$

The table of coefficients for formulae (71), (72) and (73) is given in table 3.

Table 3

Formula	a	b	c	$a - b$	$a^2 + c$	$a + b$	$a^2 - c$
(64)	$\tfrac{1}{2}r$	$\tfrac{1}{2}r$	$\tfrac{1}{4}r^2$	0	$\tfrac{1}{2}r^2$	r	0
(65)	$\tfrac{1}{2}(r - \tfrac{1}{6})$	$\tfrac{1}{2}(r + \tfrac{1}{6})$	$\tfrac{1}{4}(r + \tfrac{1}{6})^2$	$-\tfrac{1}{6}$	$\tfrac{1}{2}(r^2 + \tfrac{1}{36})$	r	$-\tfrac{1}{6}r$
(66)	r	0	r^2	r	$2r^2$	r	0

These values satisfy (76a) and (76b) for $0 \leqslant S_\beta^2$, $S_\gamma^2 \leqslant 4$, and so formulae (71), (72) and (73) are stable for all $r > 0$. Thus the A.D.I. methods obtained by splitting these formulae are unconditionally stable.

Now the stability and accuracy of split formulae depend on their equivalence with combined formulae such as (71), (72) and (73), the latter being obtained by elimination of the intermediate solution. This elimination at all internal grid points of the region $R \times [t \geqslant 0]$ is only possible if R is of rectangular type, i.e. with sides parallel to the x_1- and x_2-axes. Two examples will now be given to show how an A.D.I. method is applied when R is a rectangular type region.

Example 7

$$R = [0 \leqslant x_1, x_2 \leqslant 1]$$

The diagrams in figure 5 illustrate the grid points for $h = \tfrac{1}{4}$ at time levels (n), $(n + 1)$ and at the intermediate level $(n + 1)^*$. Before commencing the A.D.I. method using any of the split formulae mentioned, the values of $U^{n+1 *}$ are calculated at the boundary points marked X along A$'$D$'$ and B$'$C$'$. These are obtained from the values of U^n at the grid points marked X along AD and BC respectively, and the values of U^{n+1} at the grid points marked X along $A''D''$ and $B''C''$. The boundary values of U^n and U^{n+1} are of course, given. The A.D.I. method, whether D.R., D. or P.R., can now be used, solving tridiagonal sets of equations along lines parallel to the x_1- and x_2-axes at the time levels $(n+1)^*$ and $(n+1)$ respectively.

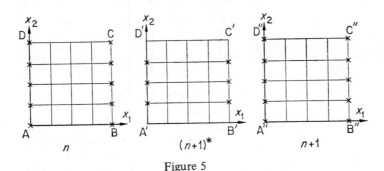

Figure 5

Example 8, $R \equiv L$-shaped region

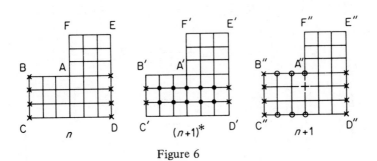

Figure 6

The reentrant corner causes trouble and the value of $U^{n+1}{}^*$ at A' cannot be calculated directly from the boundary values at time levels (n) and $(n+1)$. The following procedure is adopted. The values of $U^{n+1}{}^*$ at the grid points marked X on $B'C'$ and $E'D'$ are obtained from the values of U^n at grid points marked X along BC and ED, and the values of U^{n+1} at grid points marked X along $B''C''$ and $E''D''$. From the first step of the A.D.I. method, the values of $U^{n+1}{}^*$ are calculated at the internal grid points marked · at level $(n+1)^*$. From the second step of the A.D.I. method and the boundary values at grid points marked O at level $(n+1)$, values of U^{n+1} at internal grid points at level $(n+1)$ are calculated. These include the value of U^{n+1} at the grid point marked +, which enables the value of $U^{n+1}{}^*$ at A' to be calculated. The A.D.I. method can now proceed as in Example 7. More involved regions of rectangular type can be dealt with by repeated application of the above technique at each reentrant corner.

Exercise

16. Illustrate how an A.D.I. method can be applied when R has the following shapes

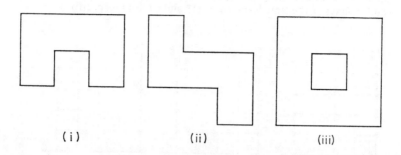

(i)   (ii)   (iii)

The analysis of A.D.I. methods, *as presented in this section*, depends on the equivalence of the combined formula and the split formula (e.g. (71) and (71a) are equivalent). *This equivalence no longer holds at grid points adjacent to boundaries of non-rectangular regions.* For example, with reference to the grid point P in figure 4, the Douglas–Rachford method (73a) becomes

and

$$
\left.
\begin{aligned}
U_{l,m}^{n+1\,*} - r &\left[\frac{2}{1+a} U_{l+1,m}^{n+1\,*} + \frac{2}{a(1+a)} U_{l-a,m}^{n+1\,*} - \frac{2}{a} U_{l,m}^{n+1\,*} \right] = \\
&U_{l,m}^{n} + r\left[\frac{2}{1+b} U_{l,m-1}^{n} + \frac{2}{b(1+b)} U_{l,m-b}^{n} - \frac{2}{b} U_{l,m}^{n} \right] \\[2ex]
U_{l,m}^{n+1} - r &\left[\frac{2}{1+b} U_{l,m-1}^{n+1} + \frac{2}{b(1+b)} U_{l,m+b}^{n} - \frac{2}{b} U_{l,m}^{n} \right] = \\
&U_{l,m}^{n+1\,*} - r\left[\frac{2}{1+b} U_{l,m-1}^{n} + \frac{2}{b(1+b)} U_{l,m+b}^{n} - \frac{2}{b} U_{l,m}^{n} \right],
\end{aligned}
\right\} \quad (73c)
$$

It is not possible to eliminate the values $U_{l,m}^{n+1\,*}$, $U_{l+1,m}^{n+1\,*}$ and $U_{l-a,m}^{n+1\,*}$ in (73c). *Nevertheless, most A.D.I. methods can still be used to solve heat conduction problems in regions of arbitrary cross-section.* The accuracy and convergence of these methods depends on the shape of the region, and a theoretical discussion of A.D.I. methods for regions of general cross-section is well beyond the scope of this book.

Example 9. Show how the Peaceman–Rachford method (71a) can be used to solve the heat conduction equation in a region with a circular cross-section of unit radius as illustrated in figure 7.

The first step of the method is given by:

$$
\begin{bmatrix} U_1 \\ U_2 \\ U_3 \\ U_4 \\ U_5 \\ U_6 \\ U_7 \\ U_8 \\ U_9 \end{bmatrix}^{(n+1)*}
=
\begin{bmatrix}
1+\frac{r}{a} & -\tfrac12 r & 0 & 0 & 0 & 0 & 0 & 0 & 0 \\
-\frac{r}{1+a} & 1+r & -\frac{r}{1+a} & 0 & 0 & 0 & 0 & 0 & 0 \\
0 & -\tfrac12 r & 1+\frac{r}{a} & 0 & 0 & 0 & 0 & 0 & 0 \\
0 & 0 & 0 & 1+\frac{r}{a} & -\tfrac12 r & 0 & 0 & 0 & 0 \\
0 & 0 & 0 & -\tfrac12 r & 1+r & -\tfrac12 r & 0 & 0 & 0 \\
0 & 0 & 0 & 0 & -\tfrac12 r & 1+r & 0 & 0 & 0 \\
0 & 0 & 0 & 0 & 0 & 0 & 1+r & -\tfrac12 r & 0 \\
0 & 0 & 0 & 0 & 0 & -\frac{r}{1+a} & -\tfrac12 r & 1+r & -\frac{r}{1+a} \\
0 & 0 & 0 & 0 & 0 & 0 & 0 & -\tfrac12 r & 1+\frac{r}{a}
\end{bmatrix}
\begin{bmatrix} U_1 \\ U_2 \\ U_3 \\ U_4 \\ U_5 \\ U_6 \\ U_7 \\ U_8 \\ U_9 \end{bmatrix}^{(n)}
$$

$$
\;+\;
\begin{bmatrix}
1-\frac{r}{a} & 0 & 0 & 0 & 0 & \tfrac12 r & 0 & 0 & 0 \\
0 & 1-r & 0 & 0 & \tfrac12 r & 0 & 0 & 0 & 0 \\
0 & 0 & \frac{r}{1+a} & 0 & 0 & \tfrac12 r & 0 & 0 & \frac{r}{1+a} \\
\frac{r}{1+a} & 0 & 0 & 1-r & 0 & 0 & \tfrac12 r & 0 & 0 \\
0 & 1-r & 0 & 0 & 1-r & 0 & 0 & \tfrac12 r & 0 \\
1-r & 0 & 0 & \tfrac12 r & 0 & \tfrac12 r & 0 & 0 & 0 \\
1-\frac{r}{a} & 0 & 0 & 0 & \tfrac12 r & 0 & 0 & \tfrac12 r & 0 \\
0 & 1-r & 0 & 0 & 0 & \tfrac12 r & 0 & 0 & 0 \\
0 & 0 & \tfrac12 r & 0 & 0 & 0 & 0 & 0 & 1-\frac{r}{a}
\end{bmatrix}
\begin{bmatrix} U_1 \\ U_2 \\ U_3 \\ U_4 \\ U_5 \\ U_6 \\ U_7 \\ U_8 \\ U_9 \end{bmatrix}^{(n)}
$$

$$
\;+\;\frac{r}{a(1+a)}
\begin{bmatrix}
U_P^{(n+1)*} + U_B^{(n)} \\[2pt]
\tfrac12 a(1+a)U_C^{(n)} \\[2pt]
U_F^{(n+1)*} + U_D^{(n)} \\[2pt]
\tfrac12 a(1+a)U_\delta^{(n+1)*} \\[2pt]
0 \\[2pt]
\tfrac12 a(1+a)U_G^{(n+1)*} \\[2pt]
U_N^{(n+1)*} + U_L^{(n)} \\[2pt]
\tfrac12 a(1+a)U_K^{(n)} \\[2pt]
U_H^{(n+1)*} + U_J^{(n)}
\end{bmatrix}
$$

68

The second step is given similarly. From figure 7, $a = \frac{1}{2}(\sqrt{3} - 1)$. The intermediate boundary conditions are taken from the original Peaceman−Rachford formulation to be

$$U_{l,m}^{n+1\,*} = g_{l,m}^{n+1/2}$$

at points on the boundary.

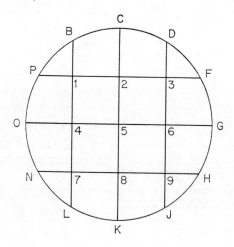

Figure 7

We now extend the application of A.D.I. methods *to the case of variable coefficients*. Although arguments based on the elimination of the intermediate solution do not always generalize to this case, most A.D.I. methods can still be applied. Consider the parabolic equation (59) with $b = c = 0$ in (60). The Peaceman−Rachford method (71a) generalizes to

$$\left.\begin{array}{l}
\left[1 - \frac{1}{2}r\delta_{x_1}(a_1\delta_{x_1})\right] U^{n+1\,*} = \left[1 + \frac{1}{2}r\delta_{x_2}(a_2\delta_{x_2})\right] U^n \\[4mm]
\text{and} \\[4mm]
\left[1 - \frac{1}{2}r\delta_{x_2}(a_2\delta_{x_2})\right] U^{n+1} = \left[1 + \frac{1}{2}r\delta_{x_1}(a_1\delta_{x_1})\right] U^{n+1\,*}
\end{array}\right\} \qquad (77)$$

for this case. In a problem where the coefficients a_1, a_2 and the boundary conditions are time dependent, the recognized procedure is to evaluate the coefficients in (77) at $t = (n + \frac{1}{2})k$, and to take the intermediate boundary values to be $U^{n+1\,*} = g_{l,m}^{n+1/2}$ for all grid points on the boundary. A more accurate boundary procedure, however, is to take

$$U^{n+1\,*} = \frac{1}{2}\left[1 - \frac{1}{2}r\delta_{x_2}(a_2\delta_{x_2})\right]g_{l,m}^{n+1} + \frac{1}{2}\left[1 + \frac{1}{2}r\delta_{x_2}(a_2\delta_{x_2})\right]g_{l,m}^{n},$$

a result which follows easily from (77). Other A.D.I. methods generalize in a similar manner.

Exercises

17. Generalize the Douglas–Rachford method (73a) to the variable coefficient case given by equations (59) and (60) with $b = c = 0$.

18. Write out the two steps of the Peaceman–Rachford method (77) for the nine grid points in the circular region of Example 8.

Another technique which can be adopted in the derivation of A.D.I. methods for variable coefficient problems, is to extend an implicit method for the case of one space variable to two or more space variables, and then to split the resulting formula into two or more parts. For example, in solving the parabolic equation

$$\frac{\partial u}{\partial t} = a(x_1, x_2, t)\frac{\partial^2 u}{\partial x_1^2} + b(x_1, x_2, t)\frac{\partial^2 u}{\partial x_2^2} \quad (a > 0, b > 0), \qquad (78)$$

a natural extension of equation (35a) to two space dimensions is

$$F = 0$$

where

$$F = (1 + X_1 - X_2)(1 + Y_1 - Y_2)U^{n+1} -$$

$$(1 + X_1 + X_2)(1 + Y_1 + Y_2)U^n$$

with

$$X_1 = \tfrac{1}{12}a\delta_{x_1}^2 a^{-1},$$

$$X_2 = \tfrac{1}{2}ra\delta_{x_1}^2,$$

$$Y_1 = \tfrac{1}{12}b\delta_{x_2}^2 b^{-1},$$

$$Y_2 = \tfrac{1}{2}rb\delta_{x_2}^2,$$

and a, b representing $a_{l,m}^{n+1/2}$, $b_{l,m}^{n+1/2}$ respectively. Expanding by the Taylor series about the point $x_1 = lh$, $x_2 = mh$, $t = (n + \tfrac{1}{2})k$, and using the differential equation to eliminate $\partial u/\partial t$, the result

$$F = -\tfrac{1}{12}rh^4\left[a\frac{\partial^2}{\partial x_1^2}\left\{b\frac{\partial^2}{\partial x_2^2}(b^{-1}u)\right\} - b\frac{\partial^2}{\partial x_2^2}\left\{b^{-1}a\frac{\partial^2 u}{\partial x_1^2}\right\}\right]_{l,m}^{n+1/2}$$

is obtained, where only terms up to and including those of fourth order in h have been retained. It follows, therefore, that the difference formula

$$(1 + X_1 - X_2)(1 + Y_1 - Y_2)U^{n+1} =$$

$$\{(1 + X_1 + X_2)(1 + Y_1 + Y_2) - 2(X_2Y_1 - Y_1X_2)\}U^n$$

is a high accuracy (i.e. locally correct to h^4) replacement of the differential equation. Perhaps a more natural form of the high accuracy difference equation is

$$(1 + X_1 - X_2)(1 + Y_1 - Y_2)(U^{n+1} - U^n) =$$

$$2(X_2 + Y_2 + X_1 Y_2 + Y_1 X_2)U^n.$$

In order to facilitate computation, this formula is split into the A.D.I. form

$$\left.\begin{array}{l} (1 + X_1 - X_2)U^{n+1\,*} = 2(X_2 + Y_2 + X_1 Y_2 + Y_1 X_2)U^n \\ (1 + Y_1 - Y_2)(U^{n+1} - U^n) = U^{n+1\,*}, \end{array}\right\} \tag{78a}$$

and

where $U^{n+1\,*}$ is an intermediate value. When R is a square, the intermediate boundary values are given by

$$g_{l,m}^{n+1\,*} = (1 + Y_1 - Y_2)(g_{l,m}^{n+1} - g_{l,m}^n). \tag{78b}$$

The stability condition for a difference scheme with variable coefficients is difficult to determine. An important paper by Widlund (1966) can be used and stability or otherwise is established in the L_2 norm. In many cases, the stability requirement can be obtained by using von Neumann's method locally at different parts of the region under consideration, the coefficients being 'frozen' to fixed values in each region. The high accuracy difference replacement derived here for the parabolic equation with variable coefficients can be shown to be unconditionally stable.

2.13 Locally one-dimensional (L.O.D.) methods

In contrast Russian numerical analysts, in particular D'Yakonov, Marchuk, Samarskii and Yanenko, have developed locally one-dimensional (L.O.D.) methods for solving time dependent partial differential equations in two or more space variables. These methods will be illustrated first with respect to the equation

$$\frac{\partial u}{\partial t} = \frac{\partial^2 u}{\partial x_1^2} + \frac{\partial^2 u}{\partial x_2^2}, \tag{79}$$

which is written as the pair of equations

$$\tfrac{1}{2}\frac{\partial u}{\partial t} = \frac{\partial^2 u}{\partial x_2^2} \tag{79a}$$

and

$$\tfrac{1}{2}\frac{\partial u}{\partial t} = \frac{\partial^2 u}{\partial x_1^2}. \tag{79b}$$

In advancing a calculation from $t = nk$ to $t = (n + 1)k$, it is assumed that equation (79a) holds from $t = nk$ to $t = (n + \frac{1}{2})k$, and equation (79b) holds from $t = (n + \frac{1}{2})k$ to $t = (n + 1)k$. The simplest explicit discretizations of these formulae are

$$U^{n+1/2} = (1 + r\delta_{x_2}^2)U^n \tag{80a}$$

and

$$U^{n+1} = (1 + r\delta_{x_1}^2)U^{n+1/2}, \tag{80b}$$

and elimination of $U^{n+1/2}$ leads to

$$U^{n+1} = (1 + r\delta_{x_1}^2)(1 + r\delta_{x_2}^2)U^n.$$

This is formula (64), which approximates equation (79) with a truncation error of $0(k^2 + kh^2)$. For a pure initial value problem with data given on the plane $t = 0$, $-\infty < x_1, x_2 < +\infty$, calculations based on the L.O.D. formulae (80a and b) have the same accuracy and stability properties but are more economical than those using formula (64). For U^{n+1} calculated at each grid point, six function evaluations are required using formulae (80a and b) as against nine using formula (64), resulting in an economy of three function evaluations at each grid point. For an initial boundary value problem in the region $R \times [t \geqslant 0]$, the intermediate boundary conditions are taken to be

$$U^{n+1/2} = g_{l,m}^{n+1/2}.$$

When R is a rectangle, however, with sides parallel to the x_1- and x_2-axes it is easily seen from (80a) that the intermediate boundary conditions should be given by

$$U^{n+1/2} = (1 + r\delta_{x_2}^2)g_{l,m}^n$$

for grid points on the boundaries parallel to the x_2-axis. Thus, for this problem, the explicit L.O.D. method consists of

$$
\left.
\begin{aligned}
U^{n+1\,*} &= (1 + r\delta_{x_2}^2)U^n \\
U^{n+1} &= (1 + r\delta_{x_1}^2)U^{n+1\,*},
\end{aligned}
\right\} \tag{81}
$$

and

with the required intermediate boundary conditions given by

$$U^{n+1\,*} = (1 + r\delta_{x_2}^2)g_{l,m}^n. \tag{82}$$

When R is an arbitrary region, the intermediate boundary conditions are usually taken to be

$$U^{n+1*} = U^{n+1/2} = g_{l,m}^{n+1/2},$$

with a subsequent loss in accuracy.

From formula (31), the Crank–Nicolson implicit discretizations of (79a and b) are

$$(1 - \tfrac{1}{2}r\delta_{x_2}^2)U^{n+1/2} = (1 + \tfrac{1}{2}r\delta_{x_2}^2)U^n$$

and

$$(1 - \tfrac{1}{2}r\delta_{x_1}^2)U^{n+1} = (1 + \tfrac{1}{2}r\delta_{x_1}^2)U^{n+1/2},$$

$$\left.\right\} \qquad (83)$$

and from formula (26), the high accuracy formulae are

$$[1 - \tfrac{1}{2}(r - \tfrac{1}{6})\delta_{x_2}^2]U^{n+1/2} = [1 + \tfrac{1}{2}(r + \tfrac{1}{6})\delta_{x_2}^2]U^n$$

and

$$[1 - \tfrac{1}{2}(r - \tfrac{1}{6})\delta_{x_1}^2]U^{n+1} = [1 + \tfrac{1}{2}(r + \tfrac{1}{6})\delta_{x_1}^2]U^{n+1/2}$$

$$\left.\right\} \qquad (84)$$

respectively. With the L.O.D. split schemes (83) and (84), $U^{n+1/2}$ can be eliminated to give formulae (71) and (72) respectively *only if the difference operators* $\delta_{x_1}^2$ *and* $\delta_{x_2}^2$ *commute*. This will happen in the present case when R is rectangular with sides parallel to the x_1- and x_2-axes. Since implicit difference methods are only applicable to initial boundary value problems, intermediate boundary conditions are required for the first steps in formulae (83) and (84) respectively. Once again, the recommended intermediate boundary values are given by

$$U^{n+1/2} = g_{l,m}^{n+1/2}.$$

Although Hubbard (1966) proves that the L.O.D. method, consisting of (83) together with the above intermediate boundary values, converges to the required solution of the heat conduction equation, it is felt that this choice of intermediate boundary values may cause a loss of accuracy in many problems. The stability of the implicit L.O.D. methods based on formulae (83) and (84) for all $r > 0$ follows immediately from the stability of formulae (31) and (32).

L.O.D. methods, both explicit and implicit, extend easily to regions of arbitrary cross-section R and to variable coefficients. In the latter case, the self-adjoint equation

$$\frac{\partial u}{\partial t} = \frac{\partial}{\partial x_1}\left(a_1 \frac{\partial u}{\partial x_1}\right) + \frac{\partial}{\partial x_2}\left(a_2 \frac{\partial u}{\partial x_2}\right) \qquad (85)$$

is written as the pair of equations

$$\tfrac{1}{2}\frac{\partial u}{\partial t} = \frac{\partial}{\partial x_2}\left(a_2 \frac{\partial u}{\partial x_2}\right) \qquad (86a)$$

and

$$\tfrac{1}{2}\frac{\partial u}{\partial t} = \frac{\partial}{\partial x_1}\left(a_1 \frac{\partial u}{\partial x_1}\right), \qquad (86b)$$

with corresponding discretizations.

Example 10. Show how the L.O.D. method (83) can be used to solve the heat conduction equation in the circular region illustrated in figure 7.

The first step of the method is given by:

$$\left[\begin{array}{ccccccccc}
1+\frac{r}{a} & 0 & 0 & -\frac{1}{2}r & 0 & 0 & 0 & 0 & 0\\[2pt]
0 & 1+r & 0 & 0 & -\frac{1}{2}r & 0 & 0 & 0 & 0\\[2pt]
0 & 0 & 1+\frac{r}{a} & 0 & 0 & -\frac{1}{2}r & 0 & 0 & 0\\[2pt]
-\frac{r}{1+a} & 0 & 0 & 1+r & 0 & 0 & -\frac{r}{1+a} & 0 & 0\\[2pt]
0 & -\frac{1}{2}r & 0 & 0 & 1+r & 0 & 0 & -\frac{1}{2}r & 0\\[2pt]
0 & 0 & -\frac{r}{1+a} & 0 & 0 & 1+r & 0 & 0 & -\frac{r}{1+a}\\[2pt]
0 & 0 & 0 & -\frac{1}{2}r & 0 & 0 & 1+\frac{r}{a} & 0 & 0\\[2pt]
0 & 0 & 0 & 0 & -\frac{1}{2}r & 0 & 0 & 1+r & 0\\[2pt]
0 & 0 & 0 & 0 & 0 & -\frac{1}{2}r & 0 & 0 & 1+\frac{r}{a}
\end{array}\right]
\begin{bmatrix}U_1\\U_2\\U_3\\U_4\\U_5\\U_6\\U_7\\U_8\\U_9\end{bmatrix}^{(n+1/2)}
=$$

$$\left[\begin{array}{ccccccccc}
1-\frac{r}{a} & 0 & 0 & \frac{1}{2}r & 0 & 0 & 0 & 0 & 0\\[2pt]
0 & 1-r & 0 & 0 & \frac{1}{2}r & 0 & 0 & 0 & 0\\[2pt]
0 & 0 & 1-\frac{r}{a} & 0 & 0 & \frac{1}{2}r & 0 & 0 & 0\\[2pt]
\frac{r}{1+a} & 0 & 0 & 1-r & 0 & 0 & \frac{r}{1+a} & 0 & 0\\[2pt]
0 & \frac{1}{2}r & 0 & 0 & 1-r & 0 & 0 & \frac{1}{2}r & 0\\[2pt]
0 & 0 & \frac{r}{1+a} & 0 & 0 & 1-r & 0 & 0 & \frac{r}{1+a}\\[2pt]
0 & 0 & 0 & \frac{1}{2}r & 0 & 0 & 1-\frac{r}{a} & 0 & 0\\[2pt]
0 & 0 & 0 & 0 & \frac{1}{2}r & 0 & 0 & 1-r & 0\\[2pt]
0 & 0 & 0 & 0 & 0 & \frac{1}{2}r & 0 & 0 & 1-\frac{r}{a}
\end{array}\right]
\begin{bmatrix}U_1\\U_2\\U_3\\U_4\\U_5\\U_6\\U_7\\U_8\\U_9\end{bmatrix}^{(n)}
+\ \frac{r}{a(1+a)}
\begin{bmatrix}
U_B^{(n+1/2)} + U_B^{(n)}\\[2pt]
\tfrac{1}{2}a(1+a)\left(U_C^{(n+1/2)} + U_C^{(n)}\right)\\[2pt]
U_D^{(n+1/2)} + U_D^{(n)}\\[2pt]
0\\[2pt]
0\\[2pt]
0\\[2pt]
U_L^{(n+1/2)} + U_L^{(n)}\\[2pt]
\tfrac{1}{2}a(1+a)\left(U_K^{(n+1/2)} + U_K^{(n)}\right)\\[2pt]
U_J^{(n+1/2)} + U_J^{(n)}
\end{bmatrix}$$

74

The second step is given similarly. The intermediate boundary conditions are taken to be

$$U^{n+1/2} = g_{l,m}^{n+1/2}$$

at points on the boundary.

A.R. Gourlay has pointed out an interesting link between the Peaceman–Rachford method (given by (71a))

$$(1 - \tfrac{1}{2}r\delta_{x_1}^2)U^{n+1/2} = (1 + \tfrac{1}{2}r\delta_{x_2}^2)U^n$$

and

$$(1 - \tfrac{1}{2}r\delta_{x_2}^2)U^{n+1} = (1 + \tfrac{1}{2}r\delta_{x_1}^2)U^{n+1/2},$$

and the locally one-dimensional method (given by (83))

$$(1 - \tfrac{1}{2}r\delta_{x_2}^2)V^{n+1/2} = (1 + \tfrac{1}{2}r\delta_{x_2}^2)V^n$$

and

$$(1 - \tfrac{1}{2}r\delta_{x_1}^2)V^{n+1} = (1 + \tfrac{1}{2}r\delta_{x_1}^2)V^{n+1/2},$$

where $U^{n+1/2}$ has been written instead of U^{n+1*} in (71a) and V has been written in place of U in (83). The link is obtained by first eliminating $V^{n+1/2}$ so that

$$(1 - \tfrac{1}{2}r\delta_{x_1}^2)V^{n+1} = (1 + \tfrac{1}{2}r\delta_{x_1}^2)(1 - \tfrac{1}{2}r\delta_{x_2}^2)^{-1}(1 + \tfrac{1}{2}r\delta_{x_2}^2)V^n$$

$$= (1 + \tfrac{1}{2}r\delta_{x_1}^2)(1 + \tfrac{1}{2}r\delta_{x_2}^2)(1 - \tfrac{1}{2}r\delta_{x_2}^2)^{-1}V^n.$$

Then we put

$$V^i = (1 - \tfrac{1}{2}r\delta_{x_2}^2)U^i \quad (i = n, n+1)$$

in the last formula which leads to

$$(1 - \tfrac{1}{2}r\delta_{x_1}^2)(1 - \tfrac{1}{2}r\delta_{x_2}^2)U^{n+1} = (1 + \tfrac{1}{2}r\delta_{x_1}^2)(1 + \tfrac{1}{2}r\delta_{x_2}^2)U^n,$$

which is the combined Peaceman–Rachford formula obtained from (71a) by eliminating the intermediate value. Thus at any stage of a calculation, the solution V^n obtained from the locally one-dimensional method is connected to the solution U^n obtained from the Peaceman–Rachford method by the formula

$$V^n = (1 - \tfrac{1}{2}r\delta_{x_2}^2)U^n.$$

This connection is exploited further in Gourlay and Mitchell (1969).

Summary

Generally speaking, explicit and implicit difference formulae are used to solve initial value and initial boundary value problems respectively. The reader may be slightly confused by the large number of implicit finite difference methods proposed for solving initial boundary value problems involving linear parabolic equations in the region $R \times [t \geqslant 0]$ where R is a closed region in (x_1, x_2) space and ∂R is its boundary.

The principal features of a problem which dictate the implicit finite difference method which it is best to use, are

(i) the nature of the coefficients in the governing parabolic equation,
(ii) the shape of the region R, and
(iii) the type of boundary condition on ∂R for $t > 0$.

Table 4 illustrates the alternatives under these three headings, with the relevant code letters in brackets. Some commonly occurring types of problem and the implicit difference methods recommended for their solution are shown in table 5. For example, problem CRD means constant coefficients, rectangular region and Dirichlet boundary conditions.

Locally one-dimensional methods have not been recommended for any of the typical problems listed in table 5. This is mainly due to the lack of experimental evidence available on L.O.D. methods. It is fairly obvious that formulae (83) and (84) and discretizations of (86) will be L.O.D. alternatives to (71a), (72a) and (77) respectively. However, it seems possible that the lack of knowledge of intermediate boundary values in L.O.D. methods may cause them to be inferior in accuracy in many problems. It should be emphasized that this is merely a conjecture on the part of the authors.

Example 11. Solve the equation

$$\frac{\partial u}{\partial t} = \frac{\partial^2 u}{\partial x_1^2} + \frac{\partial^2 u}{\partial x_2^2}$$

in the region $[0 \leqslant x_1, x_2 \leqslant 1] \times [t \geqslant 0]$, *subject to the initial condition*

$$u = \sin \pi x_1 \sin \pi x_2 \quad (0 \leqslant x_1, x_2 \leqslant 1, t = 0),$$

and the boundary conditions

$$\left. \begin{array}{l} \dfrac{\partial u}{\partial x_1} - 2u = \pi\, e^{-2\pi^2 t} \sin \pi x_2 \quad x_1 = 0 \\[3mm] \dfrac{\partial u}{\partial x_1} + 2u = -\pi\, e^{-2\pi^2 t} \sin \pi x_2 \quad x_1 = 1 \end{array} \right\} \quad 0 \leqslant x_2 \leqslant 1,\ t > 0,$$

76

$$\frac{\partial u}{\partial x_2} - 2u = \pi\,e^{-2\pi^2 t}\,\sin \pi x_1 \quad x_2 = 0$$

$$\frac{\partial u}{\partial x_2} + 2u = -\pi\,e^{-2\pi^2 t}\,\sin \pi x_1 \quad x_2 = 1$$

$$\left.\right\} \quad 0 \leqslant x_1 \leqslant 1,\ t > 0,$$

using the Peaceman–Rachford method (71a) on a high speed computer. Take $h = \frac{1}{20}$ and $r = 1$, and continue the calculation for 100 time steps.

Table 4

Coefficients	Constant (C)	Equations (39) and (60) with $a_1 = a_2 = 1, c = 0$.
	Variable (V)	Either equations (59) and (60) with $c = 0 (V_1)$ or equation (78) (V_2).
Region	Rectangular (R)	Composed of straight lines parallel to the x_1- and x_2-axes.
	Arbitrary (A)	Non-rectangular
Boundary conditions	Dirichlet (D)	u given on ∂R for $t > 0$.
	Normal derivative present (N)	Linear combination of u and $\partial u/\partial n$ given on ∂R for $t > 0$.

Table 5

Problem	Method	Intermediate boundary conditions	Remarks
CRD	(72a)	(72a)	High accuracy
V_2RD	(78a)	(78b)	High accuracy
CAD	P.R. (71a) or D.R. (73a)	P.R. $g^{n+1/2}$ D.R. g^{n+1}	Both methods must be modified at grid points adjacent to the boundary. The modifications are illustrated by Example 9 for the P.R. method and by (73c) for the D.R. method.
VAD	P.R. (77) or D.R. Exercise 15	P.R. $g^{n+1/2}$ D.R. g^{n+1}	Modifications at grid points adjacent to the boundary as in CAD with allowance for variable coefficients (see Exercise 16).

Using table 4, this example is classified as CRN. The normal derivatives in the boundary conditions are replaced in a manner similar to that used in the one space dimensional case in section 2.8. The theoretical solution to the problem is given by

$$u = e^{-2\pi^2 t} \sin \pi x_1 \sin \pi x_2.$$

The maximum modulus error (E) at a grid point after a number of time steps (P) is as shown;

P	E
1	0.000963
5	0.001021
10	0.000795
20	0.000456
40	0.000198
70	0.000102
100	0.000043

2.14 Hopscotch methods

This technique, which lies somewhere between explicit and implicit was suggested by Gordon (1965) and marketed by Gourlay (1970) under the name of *hopscotch*.

For illustration we again consider equation (59), with L independent of t, in the region $R \times [t \geqslant 0]$, where R is an arbitrary closed region in (x_1, x_2) space and ∂R is its boundary. An orthogonal grid is superimposed on R with mesh length h, and the lines of the grid intersect the boundary ∂R at the points ∂R_h. The grid points inside R are denoted by R_h.

The initial step in a hopscotch method is the subdivision of the internal grid points R_h into two disjoint subsets (blocks) K_1 and K_2 where

$$K_1 \cup K_2 = R_h.$$

The hopscotch algorithm is then defined by

$$U_{l,m}^{n+1} - U_{l,m}^n = k[\theta_{l,m}^{n+1} \ L_h \ U_{l,m}^{n+1} + \theta_{l,m}^n \ L_h \ U_{l,m}^n] \qquad (87)$$

where L_h is a discretized space operator corresponding to L, and

$$\theta_{l,m}^n = \theta_p^n = \begin{cases} 1 \\ 0 \end{cases} \text{ if } n + p \text{ is } \begin{matrix} \text{odd} \\ \text{even} \end{matrix} \qquad (88)$$

where the grid point $(lh, mh) \in K_p$ $(p = 1, 2)$. It follows that when $\theta_p^n = 1$, (87) reduces to a *fully explicit* scheme, whereas when $\theta_p^n = 0$, it reduces to a *fully implicit* scheme.

Exercises

19. Show that

$$\theta_p^{n+1} + \theta_p^n = 1, \qquad \theta_p^{n+1} \, \theta_p^n = 0 \quad (p = 1,2),$$

$$\theta_p^n + \theta_{p+1}^n = 1, \qquad \theta_{p+1}^n \, \theta_p^n = 0 \quad (p = 1).$$

20. Using (87), show that

$$U_{l,m}^{n+2} - 2U_{l,m}^{n+1} + U_{l,m}^n = 0$$

at grid points where $\theta_{l,m}^n = 0$. (N.B. This result is independent of the equation being solved.)

Now define the matrix A by the relation

$$A\mathbf{U}^n = -kL_h \, \mathbf{U}^n, \tag{89}$$

where $\mathbf{U}^n$ is the vector with components $U_{l,m}^n$, the ordering in the vector being by rows (or columns) of the grid. The diagonal matrices I_1 and I_2 are then introduced where

$$[I_1 \mathbf{U}^n]_{l,m} = \theta_{l,m}^n \, U_{l,m}^n = \theta_1^n \, U_{l,m}^n$$

and

$$[I_2 \mathbf{U}^n]_{l,m} = \theta_{l,m}^n \, U_{l,m}^n = \theta_2^n \, U_{l,m}^n$$

for $(lh, mh) \in R_h$, leading to the two step global hopscotch process

$$[I + I_2 A]\mathbf{U}^{2n+1} = [I - I_1 A]\mathbf{U}^{2n}$$

$$[I + I_1 A]\mathbf{U}^{2n+2} = [I - I_2 A]\mathbf{U}^{2n+1}, \tag{90}$$

where $I_1 + I_2 = I$, and $n = 0, 1, 2, \ldots$.

We now give three examples of the hopscotch method.

(i) *Odd-even hopscotch.* This will be illustrated with respect to $u_t = u_{xx}$ with Dirichlet boundary conditions. We choose

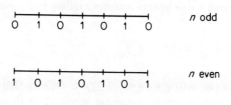

Figure 8(i)

$$K_1 = (mh, m \text{ an odd integer})$$

$$K_2 = (mh, m \text{ an even integer})$$

and so from (88),

$$\theta_m^n = \begin{matrix} 1 \\ 0 \end{matrix} \quad \text{if } n+m \text{ is } \begin{matrix} \text{odd} \\ \text{even.} \end{matrix}$$

Grid points for which θ_m^n equals unity are marked in figure 8(i) with the number 1 and those for which θ_m^n equals zero are marked with the number 0. The case illustrated has an odd number of internal grid points. Lastly we choose

$$L_h = \frac{1}{h} \, \delta_x^2.$$

The matrices are given by

$$A = -r \begin{bmatrix} -2 & 1 & & & \\ 1 & -2 & 1 & & \\ & \cdot & \cdot & \cdot & \\ & & \cdot & \cdot & \cdot \\ & & & \cdot & \cdot & \cdot \end{bmatrix}, I_1 = \begin{bmatrix} 1 & & & \\ & 0 & & \\ & & 1 & \\ & & & 0 \\ & & & & \cdot \end{bmatrix}, I_2 = \begin{bmatrix} 0 & & & \\ & 1 & & \\ & & 0 & \\ & & & 1 \\ & & & & \cdot \end{bmatrix}$$

and so the hopscotch process (90) follows.

(ii) *Line hopscotch.* This is illustrated with respect to $u_t = u_{x_1 x_1} + u_{x_2 x_2}$ with Dirichlet boundary conditions where R is a rectangle.

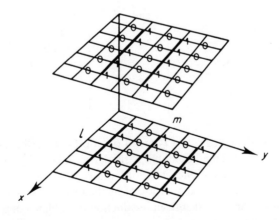

Figure 8(ii)

We select

$$K_1 = \{ (lh, mh), m \text{ an odd integer} \}$$

$$K_2 = \{ (lh, mh), m \text{ an even integer} \}.$$

Figure 8(ii) is now self explanatory, and if we choose

$$L_h = \frac{1}{h^2}(\delta_{x_1}^2 + \delta_{x_2}^2),$$

the matrices in (90) are given by

$$A = -r \begin{bmatrix} B & J & & \\ J & B & J & \\ & \cdot & \cdot & \cdot \\ & & \cdot & \cdot & \cdot \\ & & & \cdot & \cdot \end{bmatrix}, \text{ where } B = \begin{bmatrix} -4 & 1 & & \\ 1 & -4 & 1 & \\ & \cdot & \cdot & \cdot \\ & & \cdot & \cdot & \cdot \\ & & & \cdot & \cdot \end{bmatrix}, \quad J = \begin{bmatrix} 1 & & & \\ & 1 & & \\ & & \cdot & \\ & & & \cdot \end{bmatrix}$$

and

$$I_1 = \begin{bmatrix} J_1 & & & \\ & J_2 & & \\ & & J_1 & \\ & & & \cdot \\ & & & & \cdot \end{bmatrix}, \quad I_2 = \begin{bmatrix} J_2 & & & \\ & J_1 & & \\ & & J_2 & \\ & & & \cdot \\ & & & & \cdot \end{bmatrix}$$

where $J_1 = \begin{bmatrix} 1 & & & & \\ & 0 & & & \\ & & 1 & & \\ & & & 0 & \\ & & & & \cdot \\ & & & & & \cdot \end{bmatrix}$ and $J_2 = \begin{bmatrix} 0 & & & & \\ & 1 & & & \\ & & 0 & & \\ & & & 1 & \\ & & & & \cdot \\ & & & & & \cdot \end{bmatrix}.$

If $1 \leqslant m \leqslant M - 1$, then A, I_1, and I_2 are of order $(M - 1)^2$, whereas B, J, J_1, and J_2 are of order $(M - 1)$. Line hopscotch is similar to A.D.I. methods in the sense that implicit three-term recurrence relations must be solved for the unknowns. However, it should be noted that solution of implicit equations is required only on half of the grid lines. Another significant distinction is that line hopscotch is always implicit in the same direction.

Exercise

21. Calculate the local truncation errors of line hopscotch and the Peaceman–Rachford A.D.I. method for the solution of $u_t = u_{x_1 x_1} + u_{x_2 x_2}$, showing that the latter method is more accurate.

(iii) *Peripheral hopscotch.* This can be carried out on either a circular or a rectangular grid, and is particularly suited to exterior problems. In the latter case, the explicit phase of hopscotch is used to compute outwards on alternate peripherals (circles or rectangles), followed by the implicit phase on the interlaced peripherals.

The reader who wishes further details of hopscotch methods is recommended to read Gourlay (1970), Gourlay and McGuire (1971), Gane and Gourlay (1977), Gourlay and McKee (1977), and Greig and Morris (1976).

2.15 Mixed space derivative

Now consider the linear parabolic equation

$$\frac{\partial u}{\partial t} = Lu$$

where (91)

$$L \equiv a(x_1, x_2) \frac{\partial^2}{\partial x_1^2} + 2b(x_1, x_2) \frac{\partial^2}{\partial x_1 \partial x_2} + c(x_1, x_2) \frac{\partial^2}{\partial x_2^2}$$

subject to $a > 0$, $c > 0$, $b^2 < ac$ in the region $R \times [t \geqslant 0]$. The mixed partial derivative requires careful consideration and accordingly we define the difference operators

$$\sigma_1 U_{l,m}^n = U_{l+1,\,m+1}^n - U_{l,m+1}^n - U_{l+1,m}^n + U_{l,m}^n$$

$$\sigma_2 U_{l,m}^n = U_{l,m+1}^n - U_{l-1,m+1}^n - U_{l,m}^n + U_{l-1,m}^n$$

$$\sigma_3 U_{l,m}^n = U_{l,m}^n - U_{l-1,m}^n - U_{l,m-1}^n + U_{l-1,m-1}^n$$

$$\sigma_4 U_{l,m}^n = U_{l+1,m}^n - U_{l,m}^n - U_{l+1,m-1}^n + U_{l,m-1}^n.$$

Difference formulae which have been proposed for the solution of (91) include the split schemes of Samarskii (1964c)

$$(1 - ra\delta_{x_1}^2)U_{l,m}^{n+1/2} = U_{l,m}^n$$

$$(1 - rc\delta_{x_2}^2)U_{l,m}^{n+1} = (1 + rb\,(\sigma_2 + \sigma_4))\,U_{l,m}^{n+1/2}$$ (92)

and McKee and Mitchell (1970)

$$(1 - \tfrac{1}{2}ra\delta_{x_1}^2)U_{l,m}^{n+1}{}^* = (1 + \tfrac{1}{2}ra\delta_{x_1}^2 + rc\delta_{x_2}^2 + \tfrac{1}{2}rb(\sigma_1 + \sigma_2 + \sigma_3 + \sigma_4))U_{l,m}^n$$

$$(1 - \tfrac{1}{2}rc\delta_{x_2}^2)U_{l,m}^{n+1} = U_{l,m}^{n+1}{}^* - \tfrac{1}{2}rc\delta_{x_2}^2 U_{l,m}^n. \tag{93}$$

One merit of (93) is that intermediate boundary values are given by

$$U_{l,m}^{n+1}{}^* = (1 - \tfrac{1}{2}rc\delta_{x_2}^2)g_{l,m}^{n+1} + \tfrac{1}{2}rc\delta_{x_2}^2 g_{l,m}^n \tag{94a}$$

whereas (92) has to be content with

$$U_{l,m}^{n+1/2} = g_{l,m}^{n+1/2}. \tag{94b}$$

Here g is written for U when the grid point is on the boundary of the region.

Exercise

22. Eliminate $U_{l,m}^{n+1}{}^*$ in (93) and so obtain a combined formula. Show that this combined formula has a principal truncation error given by

$$\left[-\tfrac{1}{12}ra\frac{\partial^4 u}{\partial x_1^4} + (\tfrac{1}{3} + ra)rb\frac{\partial^4 u}{\partial x_1^3 \partial x_2} + 2r^2 b^2 \frac{\partial^4 u}{\partial x_1^2 \partial x_2^2} + (\tfrac{1}{3} + rc)rb\frac{\partial^4 u}{\partial x_1 \partial x_2^3} \right.$$
$$\left. - \tfrac{1}{12}rc\frac{\partial^4 u}{\partial x_2^4} \right]h^4$$

if the coefficients a, b, and c are constant. Show also in this case that the formula is unconditionally stable.

A line hopscotch scheme of Gourlay and McKee (1977) proposed for (91) is claimed to be more accurate than either of the A.D.I. schemes given by (92) and (93).

2.16 Parabolic equations in three space dimensions

Here we examine finite difference methods of solution of the equation

$$\frac{\partial u}{\partial t} = Lu, \tag{95}$$

where

$$L \equiv \sum_{i=1}^{3} \left[\frac{\partial}{\partial x_i}\left\{ a_i(x_1, x_2, x_3, t)\frac{\partial}{\partial x_i} \right\} + b_i(x_1, x_2, x_3, t)\frac{\partial}{\partial x_i} \right] - c(x_1, x_2, x_3, t), \tag{96}$$

with $a_i > 0$ $(i = 1, 2, 3)$ and $c \geq 0$. The region to be examined in (x_1, x_2, x_3, t) space is covered by a rectilinear grid with sides parallel to the axes, and h and k are the grid spacings in the space and time directions respectively. The function satisfying the difference equation at the grid point $X_1 = jh$, $X_2 = lh$, $X_3 = mh$, $T = nk$, where j, l, m, n are integers, is given by $U_{j,l,m}^n$. The exact difference replacement of equation (95) involving two adjacent time levels, provided L is independent of t, is given by

$$U_{j,l,m}^{n+1} = \exp(kL)U_{j,l,m}^n, \tag{97}$$

with

$$\frac{\partial}{\partial x_i} = \frac{2}{h} \sinh^{-1} \frac{\delta_{x_i}}{2} \ (i = 1, 2, 3),$$

where

$$\delta_{x_1} U_{j,l,m}^n = U_{j+1/2,l,m}^n - U_{j-1/2,l,m}^n.$$

There are similar expressions for $\delta_{x_2} U_{j,l,m}^n$ and $\delta_{x_3} U_{j,l,m}^n$. In the following sections, U^n will be written in short for $U_{j,l,m}^n$.

2.17 Explicit methods

It is a simple matter to extend to three space dimensions the explicit difference methods in two space dimensions given previously. Most of the details will be omitted and only the main results will be quoted. For the simple form of the operator

$$L \equiv \sum_{i=1}^{3} \frac{\partial}{\partial x_i^2},$$

equation (97) gives

$$U^{n+1} = \prod_{i=1}^{3} \left[1 + r\delta_{x_i}^2 + \tfrac{1}{2}r(r - \tfrac{1}{6})\delta_{x_i}^4 + \ldots \right] U^n. \tag{98}$$

Various explicit difference formulae can be obtained by approximating equation (98). Typical examples are

$$U^{n+1} = (1 + r\Sigma\delta_{x_i}^2)U^n + 0(k^2 + kh^2) \tag{99}$$

and

$$U^{n+1} = \prod_{i=1}^{3} (1 + r\delta_{x_i}^2)U^n + 0(k^2 + kh^2). \tag{100}$$

The von Neumann method of examining the stability of difference schemes in three space dimensions depends on the growth (or otherwise) with time of the term

$$e^{\alpha t}\, e^{i\beta x_1}\, e^{i\gamma x_2}\, e^{i\delta x_3}$$

where β, γ, δ are arbitrary real numbers and $\alpha \equiv \alpha(\beta, \gamma, \delta)$ is in general complex. The condition for stability is

$$|e^{\alpha k}| \leqslant 1$$

for all α.

Exercise

23. Show that formula (99) is stable if $0 < r \leqslant \frac{1}{6}$ and that formula (100) is stable if $0 < r \leqslant \frac{1}{2}$.

The explicit difference formula (100) involves twenty-seven grid points at the original time level $t = nk$. It is more economical from the computational point of view if it is used in the split form

$$\left. \begin{aligned} U^{n+1 *} &= (1 + r\delta_{x_3}^2)U^n \\[2mm] U^{n+1 **} &= (1 + r\delta_{x_2}^2)U^{n+1 *} \\[2mm] U^{n+1} &= (1 + r\delta_{x_1}^2)U^{n+1 **}. \end{aligned} \right\} \tag{101}$$

It is easy to see that by elimination of the intermediate solutions $U^{n+1 *}$ and $U^{n+1 **}$, formula (100) is recovered. For a pure initial value problem, the split formula (101) can be used in a straightforward manner. For an initial boundary value problem in the region $R \times [t \geqslant 0]$, however, care must be taken with the boundary conditions at the intermediate levels. For example, if R is a rectangular parallelepiped with its edges parallel to the space axes, the intermediate boundary conditions are given by

$$U^{n+1 *} = (1 + r\delta_{x_3}^2)g^n$$

and

$$U^{n+1 **} = (1 + r\delta_{x_2}^2)(1 + r\delta_{x_3}^2)g^n,$$

where

$$u(x_1, x_2, x_3, t) = g(x_1, x_2, x_3, t)$$

when (x_1, x_2, x_3, t) is on the boundary of the region. If R is arbitrary, the intermediate boundary conditions are usually taken to be

$$U^{n+1}* \equiv U^{n+1/3} = g^{n+1/3}$$

and

$$U^{n+1}** \equiv U^{n+2/3} = g^{n+2/3},$$

with a subsequent loss in accuracy. The split formula (101) with $U^{n+1}*$ and $U^{n+1}**$ replaced by $U^{n+1/3}$ and $U^{n+2/3}$ respectively can, of course, be obtained directly from the system of differential equations

$$
\left.
\begin{array}{l}
\tfrac{1}{3} \dfrac{\partial u}{\partial t} = \dfrac{\partial^2 u}{\partial x_3^2} \\[3mm]
\tfrac{1}{3} \dfrac{\partial u}{\partial t} = \dfrac{\partial^2 u}{\partial x_2^2} \\[3mm]
\tfrac{1}{3} \dfrac{\partial u}{\partial t} = \dfrac{\partial^2 u}{\partial x_1^2},
\end{array}
\right\}
\tag{102}
$$

where in advancing a calculation from $t = nk$ to $t = (n + 1)k$ it is assumed that the first equation in (102) holds from $t = nk$ to $t = (n + \tfrac{1}{3})k$, the second equation from $t = (n + \tfrac{1}{3})k$ to $t = (n + \tfrac{2}{3})k$, and the final equation from $t = (n + \tfrac{2}{3})k$ to $t = (n + 1)k$.

When the operator in equation (95) has the form

$$L \equiv \sum_{i=1}^{3} \frac{\partial}{\partial x_i} \left\{ a_i(x_1, x_2, x_3) \frac{\partial}{\partial x_i} \right\},$$

explicit formulae corresponding to (99) and (100) are given by

$$U^{n+1} = \left[1 + r \sum_{i=1}^{3} \delta_{x_i}(a_i \delta_{x_i}) \right] U^n \tag{103}$$

and

$$U^{n+1} = \prod_{i=1}^{3} (1 + r\delta_{x_i}(a_i \delta_{x_i})) U^n \tag{104}$$

respectively. Equation (104) can also be written in the split form, which can be obtained directly from the self-adjoint form of the equations (102).

2.18 Implicit methods

Explicit difference methods are rarely used to solve *initial boundary value*

D

86

problems in three space dimensions. Once again A.D.I. and L.O.D. methods are used extensively and these will now be described. In the first instance the operator

$$L \equiv \sum_{i=1}^{3} \frac{\partial}{\partial x_i^2}$$

will be considered and the solution of equation (95) will be required in the region of (x_1, x_2, x_3, t) space consisting of $R \times [t \geqslant 0]$, *where R is the cube* $[0 \leqslant x_1, x_2, x_3 \leqslant 1]$. The initial condition is

$$u(x_1, x_2, x_3, 0) = f(x_1, x_2, x_3), \quad (x_1, x_2, x_3) \in R,$$

and the boundary condition is

$$u(x_1, x_2, x_3, t) = g(x_1, x_2, x_3, t) \quad (x_1, x_2, x_3, t) \in \partial R \times [t \geqslant 0],$$

where ∂R is the boundary of R.

The *Douglas–Rachford* method (73a) generalizes easily to three space dimensions to give

$$\left.\begin{aligned}
(1 - r\delta_{x_1}^2)U^{n+1*} &= [1 + r(\delta_{x_2}^2 + \delta_{x_3}^2)]U^n \\
(1 - r\delta_{x_2}^2)U^{n+1**} &= U^{n+1*} - r\delta_{x_2}^2 U^n \\
(1 - r\delta_{x_3}^2)U^{n+1} &= U^{n+1**} - r\delta_{x_3}^2 U^n.
\end{aligned}\right\} \quad (105a)$$

Elimination of the intermediate values U^{n+1*} and U^{n+1**} gives

$$\prod_{i=1}^{3}(1 - r\delta_{xi}^2)U^{n+1} = \left[\prod_{i=1}^{3}(1 - r\delta_{xi}^2) + r\sum_{i=1}^{3}\delta_{xi}^2\right]U^n$$

$$(105)$$

The intermediate boundary values proposed for the D.R. scheme are

$$U^{n+1*} = U^{n+1**} = g^{n+1}. \quad (106)$$

In fact, it is easily seen from (105a) that they should be

$$\left.\begin{aligned}
U^{n+1*} &= g^n + (1 - r\delta_{x_2}^2)(1 - r\delta_{x_3}^2)(g^{n+1} - g^n) \\
U^{n+1**} &= g^n + (1 - r\delta_{x_3}^2)(g^{n+1} - g^n).
\end{aligned}\right\} \quad (107)$$

Exercises

24. Derive the boundary values (107) from (105a) and show that they reduce to (106) only for a problem with boundary conditions independent of the time.

25. Show that equation (105) has a truncation error $0(k^2 + kh^2)$.

An A.D.I. method suggested by *Douglas* (1962) is given by

$$\left.\begin{array}{l} (1 - \alpha\delta_{x_1}^2)U^{n+1}* = \left[\dfrac{\alpha}{\beta}(1 + \beta\delta_{x_1}^2) + (\alpha + \beta)(\delta_{x_2}^2 + \delta_{x_3}^2)\right]U^n \\[2mm] (1 - \alpha\delta_{x_2}^2)U^{n+1}** = U^{n+1}* - \beta\delta_{x_2}^2 U^n \\[2mm] (1 - \alpha\delta_{x_3}^2)U^{n+1} = \dfrac{\beta}{\alpha}U^{n+1}** - \dfrac{\beta^2}{\alpha}\delta_{x_3}^2 U^n, \end{array}\right\} \quad (108a)$$

with $\alpha = \beta = \frac{1}{2}r$. Elimination of the intermediate solution leads to

$$\prod_{i=1}^{3}(1 - \alpha\delta_{x_i}^2)U^{n+1} = \left[\prod_{i=1}^{3}(1 + \beta\delta_{x_i}^2) - (\alpha + \beta)\beta^2\prod_{i=1}^{3}\delta_{x_i}^2\right]U^n. \quad (108)$$

The intermediate boundary values proposed by Douglas are

$$U^{n+1}* = U^{n+1}** = g^{n+1}, \quad (109)$$

whereas from (108a), they are obtained as

$$\left.\begin{array}{l} U^{n+1}* = \dfrac{\beta}{\alpha}g^n + (1 - \alpha\delta_{x_2}^2)(1 - \alpha\delta_{x_3}^2)\left(\dfrac{\alpha}{\beta}g^{n+1} - \dfrac{\beta}{\alpha}g^n\right) \\[4mm] U^{n+1}** = \dfrac{\beta}{\alpha}g^n + (1 - \alpha\delta_{x_3}^2)\left(\dfrac{\alpha}{\beta}g^{n+1} - \dfrac{\beta}{\alpha}g^n\right). \end{array}\right\} \quad (110)$$

and

Again (110) reduces to (109) only if the boundary conditions are time independent. Formula (108) has a truncation error $0(k^3 + kh^2)$ when $\alpha = \beta = \frac{1}{2}r$. If $\alpha = \frac{1}{2}(r - \frac{1}{6})$, $\beta = \frac{1}{2}(r + \frac{1}{6})$, then formula (108) has a truncation error $0(k^3 + kh^4)$. In this case the intermediate boundary values (109) lead to a loss of accuracy in both the time dependent and time independent cases.

Exercise

26. With $\alpha = \beta = \frac{1}{2}r$, write down the counterpart of (108a) in two space dimensions and show that elimination of $U^{n+1}*$ leads to formula (71).

Formulae (71) and (72) can be extended to three space dimensions in the form

$$\prod_{i=1}^{3}(1 - \alpha\delta_{x_i}^2)U^{n+1} = \prod_{i=1}^{3}(1 + \beta\delta_{x_i}^2)U^n. \quad (111)$$

It is easily verified that formula (111) has a truncation error $0(k^3 + kh^2)$ when

$\alpha = \beta = \frac{1}{2}r$ and a truncation error $0(k^3 + kh^4)$ when $\alpha = \frac{1}{2}(r - \frac{1}{6}), \beta = \frac{1}{2}(r + \frac{1}{6})$. Formula (111) can be split according to a general method of splitting operators suggested by D'Yakonov, as

$$
\left.
\begin{aligned}
(1 - \alpha\delta^2_{x_1})U^{n+1}* &= \prod_{i=1}^{3} (1 + \beta\delta^2_{x_i})U^n \\[2mm]
(1 - \alpha\delta^2_{x_2})U^{n+1}** &= U^{n+1}* \\[2mm]
(1 - \alpha\delta^2_{x_3})U^{n+1} &= U^{n+1}**,
\end{aligned}
\right\} \tag{111a}
$$

or more economically in the form

$$
\left.
\begin{aligned}
(1 - \alpha\delta^2_{x_1})U^{n+1}* &= (1 + \beta\delta^2_{x_2})(1 + \beta\delta^2_{x_3})U^n \\[2mm]
(1 - \alpha\delta^2_{x_2})U^{n+1}** &= U^{n+1}* - \frac{\beta}{\alpha}(1 + \beta\delta^2_{x_3})U^n \\[2mm]
(1 - \alpha\delta^2_{x_3})U^{n+1} &= \left(1 + \frac{\beta}{\alpha}\right)U^{n+1}** + \left(\frac{\beta}{\alpha}\right)^2(1 + \beta\delta^2_{x_3})U^n.
\end{aligned}
\right\} \tag{111b}
$$

The intermediate boundary values obtained from (111b) are

$$
U^{n+1}* = \frac{\alpha}{\alpha+\beta}(1 - \alpha\delta^2_{x_2})(1 - \alpha\delta^2_{x_3})g^{n+1} + \frac{\beta}{\alpha+\beta}(1 + \beta\delta^2_{x_2})(1 + \beta\delta^2_{x_3})g^n
$$

and

$$
U^{n+1}** = \frac{\alpha}{\alpha+\beta}(1 - \alpha\delta^2_{x_3})g^{n+1} - \frac{\alpha}{\alpha+\beta}\left(\frac{\beta}{\alpha}\right)^2(1 + \beta\delta^2_{x_3})g^n
$$

respectively.

The methods given by the split formulae (105a), (108a), (111a) and (111b) are all three-step methods involving the solution of tridiagonal sets of equations along lines parallel to the x_1-, x_2-, and x_3-axes at the first, second and third steps respectively, and so are classified as A.D.I. methods. It is easily shown using von Neumann's method of examining stability that all the formulae mentioned in this section are unconditionally stable except (108), with $\alpha = \frac{1}{2}(r - \frac{1}{6})$, $\beta = \frac{1}{2}(r + \frac{1}{6})$, which is stable only if $r \leqslant \frac{11}{12}$.

Exercise
27. Verify that formulae (105) and (111) are unconditionally stable.

Finally, from formula (102), we get the locally one-dimensional scheme

$$
\left.
\begin{aligned}
(1 - \alpha\delta_{x_3}^2)U^{n+1/3} &= (1 + \beta\delta_{x_3}^2)U^n \\
(1 - \alpha\delta_{x_2}^2)U^{n+2/3} &= (1 + \beta\delta_{x_2}^2)U^{n+1/3} \\
(1 - \alpha\delta_{x_1}^2)U^{n+1} &= (1 + \beta\delta_{x_1}^2)U^{n+2/3},
\end{aligned}
\right\}
\tag{112}
$$

where $\alpha = \beta = \tfrac{1}{2}r$, or in the high accuracy case $\alpha = \tfrac{1}{2}(r - \tfrac{1}{6})$, $\beta = \tfrac{1}{2}(r + \tfrac{1}{6})$. With suitable commutation of the operators, $U^{n+1/3}$ and $U^{n+2/3}$ can be eliminated from scheme (112) to give formula (111). Again the recommended boundary values are given by

$$
U^{n+1/3} = g^{n+1/3}
$$

and

$$
U^{n+2/3} = g^{n+2/3},
$$

which will cause a loss of accuracy in many problems, particularly in the high accuracy case where $\alpha = \tfrac{1}{2}(r - \tfrac{1}{6})$, $\beta = \tfrac{1}{2}(r + \tfrac{1}{6})$.

The modifications required in the A.D.I. and L.O.D. methods in three space dimensions, to cater for variable coefficients and arbitrary regions, follow a similar pattern to those required in two space dimensions and will not be repeated here.

2.19 Three-level difference schemes

It may be advantageous in the construction of difference schemes to use more time levels than the minimum number required by the differential equation. The extra level (or levels) is usually introduced in order to produce a scheme of higher accuracy or, occasionally, improved stability. In the present section, we shall restrict ourselves to the consideration of *three-level* difference replacements of parabolic equations. When a three-level scheme is used, initial data is required on $t = 0$ and $t = k$ ($n = 0$ and 1) to start the calculation. Since only data at $t = 0$ is given, it is necessary to calculate the data at $t = k$ from a two-level difference formula or by using a power series expansion. The data calculated at $t = k$ should be of an accuracy comparable with that of the three-level scheme.

2.20 Explicit schemes

The simplest three-level explicit scheme for the solution of the diffusion equation

$$
\frac{\partial u}{\partial t} = \frac{\partial^2 u}{\partial x^2}
$$

is

$$\frac{1}{2k}(U_m^{n+1} - U_m^{n-1}) = \frac{1}{h^2}(U_{m+1}^n - 2U_m^n + U_{m-1}^n),$$

which can be written in the form

$$U_m^{n+1} = 2r\delta_x^2 U_m^n + U_m^{n-1}. \tag{113}$$

The stability of formula (113) can be examined by writing it as the two-level system

$$U_m^{n+1} = 2r\delta_x^2 U_m^n + V_m^n$$

and

$$V_m^{n+1} = U_m^n, \tag{114}$$

which can be written in vector form as

$$\begin{bmatrix} U \\ V \end{bmatrix}_m^{n+1} = \begin{bmatrix} 2r\delta_x^2 & 1 \\ 1 & 0 \end{bmatrix} \begin{bmatrix} U \\ V \end{bmatrix}_m^n.$$

If we introduce the vector $W = \begin{bmatrix} U \\ V \end{bmatrix}$, and a typical Fourier term

$$W_m^n = W_0^n\, e^{i\beta mh},$$

where W_0^n is a constant vector, is substituted into (114), the result

$$W_m^{n+1} = \begin{bmatrix} -8r\sin^2\frac{\beta h}{2} & 1 \\ 1 & 0 \end{bmatrix} W_m^n$$

is obtained. The above matrix is called the *amplification matrix* of the system (see Chapter 4, Hyperbolic Equations) and its eigenvalues $\mu_i(i = 1, 2)$ are given by

$$\mu_i = -4r\sin^2\frac{\beta h}{2} \pm \left(1 + 16r^2\sin^4\frac{\beta h}{2}\right)^{1/2}.$$

Now the von Neumann *necessary* condition for the stability of a two-level system is

$$\max |\mu_i| \leqslant 1 \quad (i = 1, 2),$$

where $\mu_i(i = 1, 2)$ are the eigenvalues of the amplification matrix. This condition is violated by μ_2 for all values of r, and so the method (113) is unconditionally unstable.

Du Fort and Frankel (1953) proposed that $\delta_x^2 U_m^n$ be rewritten as $(U_{m+1}^n - 2U_m^n + U_{m-1}^n)$ in the scheme (113) and that U_m^n be then replaced by $\frac{1}{2}(U_m^{n+1} + U_m^{n-1})$, leading to the scheme

$$(1 + 2r)U_m^{n+1} = 2r(U_{m+1}^n + U_{m-1}^n) + (1 - 2r)U_m^{n-1}. \tag{115}$$

This scheme, although still explicit, can be shown to be stable for all r.

Exercises

28. Replace scheme (115) by a two-level system and show that the eigenvalues of the amplification matrix are given by

$$\mu_i = \frac{2r\cos\beta h \pm (1 - 4r^2\sin^2\beta h)^{1/2}}{1 + 2r} \quad (i = 1, 2).$$

Hence show that the Du Fort–Frankel scheme satisfies the von Neumann condition necessary for stability.

29. Show that the principal part of the truncation error of the Du Fort–Frankel scheme is

$$2k\left[\frac{k^2}{h^2}\frac{\partial^2 u}{\partial t^2} - \frac{1}{12}h^2\frac{\partial^4 u}{\partial x^4}\right]_m^n.$$

Now the condition of consistency for a parabolic equation is that

$$\frac{\text{Truncation error}}{k} \to 0 \text{ as } h, k \to 0.$$

For the Du Fort–Frankel scheme, the consistency condition is satisfied only if $k/h \to 0$ with the grid spacing. If r is kept constant, as $h, k \to 0$, then

$$\frac{k}{h} = rh \propto h,$$

and so the Du Fort–Frankel scheme is consistent. If, however,

$$\frac{k}{h} = c = \text{a fixed constant},$$

then the Du Fort–Frankel scheme is consistent with the hyperbolic equation

$$\frac{\partial u}{\partial t} + c^2\frac{\partial^2 u}{\partial t^2} = \frac{\partial^2 u}{\partial x^2}.$$

Exercise

30. Show that the formula of maximum accuracy involving the grid points

92

$(mh, (n + 1)k), (mh, nk), (mh, (n - 1)k), ((m + 1)h, nk),$ and $((m - 1)h, nk)$ is

$$(1 + 6r)U_m^{n+1} = 2(1 - 12r^2)U_m^n + 12r^2(U_{m+1}^n + U_{m-1}^n) - (1 - 6r)U_m^{n-1}.$$

Find the truncation error of this formula and show that it is stable provided that $r \leqslant 1/(2\sqrt{3})$.

2.21 Implicit schemes

In general, implicit schemes have better stability properties than explicit schemes. If we consider the unconditionally unstable scheme (113) and replace $\delta_x^2 U_m^n$ by $\frac{1}{3}\delta_x^2(U_m^{n+1} + U_m^n + U_m^{n-1})$, the implicit scheme

$$(1 - \tfrac{2}{3}r\delta_x^2)U_m^{n+1} = \tfrac{2}{3}r\delta_x^2 U_m^n + (1 + \tfrac{2}{3}r\delta_x^2)U_m^{n-1} \tag{116}$$

is obtained. This time the roots of the amplification matrix are given by

$$(1 + a)\mu^2 + a\mu - (1 - a) = 0,$$

where

$$a = \tfrac{8}{3}r \sin^2 \frac{\beta h}{2},$$

and so

$$\mu = \frac{-a \pm (4 - 3a^2)^{1/2}}{2(1 + a)}.$$

It is easy to see that $|\mu| \leqslant 1$ for all a, and so scheme (116) is unconditionally stable in the sense of von Neumann.

For the diffusion equation

$$\frac{\partial u}{\partial t} = \frac{\partial^2 u}{\partial x_1^2} + \frac{\partial^2 u}{\partial x_2^2},$$

in two space dimensions, the unconditionally stable formula

$$[1 - \tfrac{2}{3}r(\delta_{x_1}^2 + \delta_{x_2}^2)] U^{n+1} = \tfrac{2}{3}r(\delta_{x_1}^2 + \delta_{x_2}^2)U^n + [1 + \tfrac{2}{3}r(\delta_{x_1}^2 + \delta_{x_2}^2)] U^{n-1} \tag{117}$$

is constructed similarly to scheme (116). In order to obtain an operator on the left-hand side of formula (117) which factorizes, the fourth-order term $\tfrac{4}{9}r^2\delta_{x_1}^2\delta_{x_2}^2 U^{n+1}$ is added, leading to

$$(1 - \tfrac{2}{3}r\delta_{x_1}^2)(1 - \tfrac{2}{3}r\delta_{x_2}^2)U^{n+1} = [\tfrac{2}{3}r(\delta_{x_1}^2 + \delta_{x_2}^2) + \tfrac{4}{9}r^2\alpha\delta_{x_1}^2\delta_{x_2}^2] U^n +$$
$$[1 + \tfrac{2}{3}r(\delta_{x_1}^2 + \delta_{x_2}^2) + \tfrac{4}{9}r^2\beta\delta_{x_1}^2\delta_{x_2}^2] U^{n-1}, \tag{118}$$

where other fourth-order terms involving parameters α and β have been included. The fourth-order terms do not alter the order of accuracy but can have a decided influence on the structure of the split formulae which come from (118). They also influence the stability of the composite formula. A typical splitting of formula (118) is

$$\left.\begin{aligned}
(1 - \tfrac{2}{3}r\delta_{x_1}^2)U^{n+1}* &= \tfrac{2}{3}r\delta_{x_2}^2(\alpha U^n + \beta U^{n-1}) + \tfrac{2}{3}r(\delta_{x_1}^2 + \delta_{x_2}^2)U^n + \\
[1 &+ \tfrac{2}{3}r(\delta_{x_1}^2 + \delta_{x_2}^2)]U^{n-1}
\end{aligned}\right\} (119)$$

and

$$(1 - \tfrac{2}{3}r\delta_{x_2}^2)U^{n+1} = U^{n+1}* - \tfrac{2}{3}r\delta_{x_2}^2(\alpha U^n + \beta U^{n-1}),$$

and two special cases are worth mentioning:

(i) $\alpha + \beta = 1$, leading to the first formula in (119) having the same order of accuracy as the composite formula (118).

(ii) $\alpha = \beta = 0$, making (119) equivalent to the D'Yakonov splitting, already mentioned for two-level schemes.

Example 12 Show that only when $\alpha + \beta = 1$, is the order of accuracy of the first formula in (119) equivalent to the order of accuracy of the composite formula (118).

Using the relationships

$$\delta_{x_i}^2 U \approx h^2 \frac{\partial^2 u}{\partial x_{x_i}^2} + \tfrac{1}{12}h^4 \frac{\partial^4 u}{\partial x_{x_i}^4}, \qquad (i = 1, 2)$$

$$U^{n+1} \approx u + k \frac{\partial u}{\partial t} + \tfrac{1}{2}k^2 \frac{\partial^2 u}{\partial t^2},$$

$$U^{n-1} \approx u - k \frac{\partial u}{\partial t} + \tfrac{1}{2}k^2 \frac{\partial^2 u}{\partial t^2},$$

it is easily shown that subtraction of the right-hand side of formula (118) from the left-hand side leads to

$$2k\left(\frac{\partial u}{\partial t} - \frac{\partial^2 u}{\partial x_1^2} - \frac{\partial^2 u}{\partial x_2^2}\right) - \tfrac{1}{6}kh^2\left(\frac{\partial^4 u}{\partial x_1^4} + \frac{\partial^4 u}{\partial x_2^4}\right) + \tfrac{4}{9}k^2\left\{1 - (\alpha + \beta)\right\}\frac{\partial^4 u}{\partial x_1^2 \partial x_2^2}.$$

Similarly from the first formula of (119) we get

$$2k\left(\frac{\partial u}{\partial t} - \frac{\partial^2 u}{\partial x_1^2} - \frac{2 + \alpha + \beta}{3}\frac{\partial^2 u}{\partial x_2^2}\right) - \tfrac{1}{6}kh^2\left(\frac{\partial^4 u}{\partial x_1^4} + \frac{2 + \alpha + \beta}{3}\frac{\partial^4 u}{\partial x_2^4}\right) +$$

$$\tfrac{2}{3}k^2(\beta+1)\ \frac{\partial^3 u}{\partial x_2^2 \partial t}.$$

These formulae agree to terms of second order if

$$\alpha + \beta = 1.$$

If in addition,

$$\beta + 1 = 0,$$

giving $\beta = -1$ and $\alpha = 2$, the formulae agree to terms of fourth order.

A higher accuracy replacement of the diffusion equation in two space dimensions is given by

$$[1 + (\tfrac{1}{12} - \tfrac{2}{3}r)(\delta_{x_1}^2 + \delta_{x_2}^2)]\,U^{n+1} = \tfrac{2}{3}r[\delta_{x_1}^2 + \delta_{x_2}^2 + \tfrac{1}{2}\delta_{x_1}^2\delta_{x_2}^2]\,U^n +$$

$$[1 + (\tfrac{1}{12} + \tfrac{2}{3}r)(\delta_{x_1}^2 + \delta_{x_2}^2)]\,U^{n-1}. \qquad (120)$$

In order to obtain an operator on the left-hand side which factorizes, the fourth-order term $(\tfrac{1}{12} - \tfrac{2}{3}r)^2 \delta_{x_1}^2 \delta_{x_2}^2$ is added, leading to

$$[1 + (\tfrac{1}{12} - \tfrac{2}{3}r)\delta_{x_1}^2\,]\,[1 + (\tfrac{1}{12} - \tfrac{2}{3}r)\delta_{x_2}^2\,]\,U^{n+1} = \tfrac{2}{3}r[\delta_{x_1}^2 + \delta_{x_2}^2 + \tfrac{1}{2}\delta_{x_1}^2\delta_{x_2}^2]\,U^n$$

$$+ [1 + (\tfrac{1}{12} + \tfrac{2}{3}r)(\delta_{x_1}^2 + \delta_{x_2}^2)]\,U^{n-1} + (\tfrac{1}{12} - \tfrac{2}{3}r)^2\delta_{x_1}^2\delta_{x_2}^2(\alpha U^n + \beta U^{n-1}), \quad (121)$$

where other fourth-order terms involving parameters α and β have been included. To maintain the order of accuracy of the original formula, we require

$$\alpha + \beta = 1.$$

A useful looking solution is

$$\alpha = -\ \frac{\tfrac{2}{3}r}{(\tfrac{1}{12} - \tfrac{2}{3}r)^2}\ , \qquad \beta = \frac{(\tfrac{1}{12} + \tfrac{2}{3}r)^2}{(\tfrac{1}{12} - \tfrac{2}{3}r)^2}\ ,$$

leading to the formula

$$[1 + (\tfrac{1}{12} - \tfrac{2}{3}r)\delta_{x_1}^2\,]\,[1 + (\tfrac{1}{12} - \tfrac{2}{3}r)\delta_{x_2}^2\,]\,U^{n+1} = \tfrac{2}{3}r[\delta_{x_1}^2 + \delta_{x_2}^2 + \tfrac{1}{6}\delta_{x_1}^2\delta_{x_2}^2]\,U^n$$

$$+ [1 + (\tfrac{1}{12} + \tfrac{2}{3}r)\delta_{x_1}^2\,]\,[1 + (\tfrac{1}{12} + \tfrac{2}{3}r)\delta_{x_2}^2\,]\,U^{n-1}. \qquad (122)$$

Exercises

31. Show that equation (120) is a high accuracy difference replacement of the diffusion equation in two space dimensions.

32. Examine the stability of formula (122) in the sense of von Neumann.

It is instructive to detail the steps of the stability analysis for formula (122). The roots of the amplification matrix are μ_i $(i = 1, 2)$ given by

$$\left[1 + (\tfrac{1}{12} - \tfrac{2}{3}r)\left(-4\sin^2\frac{\beta h}{2}\right)\right]\left[1 + (\tfrac{1}{12} - \tfrac{2}{3}r)\left(-4\sin^2\frac{\gamma h}{2}\right)\right]\mu^2 +$$

$$\tfrac{2}{3}r\left[-4\sin^2\frac{\beta h}{2} - 4\sin^2\frac{\gamma h}{2} + \tfrac{8}{3}\sin^2\frac{\beta h}{2}\sin^2\frac{\gamma h}{2}\right]\mu -$$

$$\left[1 + (\tfrac{1}{12} + \tfrac{2}{3}r)\left(-4\sin^2\frac{\beta h}{2}\right)\right]\left[1 + (\tfrac{1}{12} + \tfrac{2}{3}r)\left(-4\sin^2\frac{\gamma h}{2}\right)\right] = 0,$$

which can be written more concisely as

$$[(1 - \tfrac{1}{3}a) + \tfrac{8}{3}ra]\,[(1 - \tfrac{1}{3}b) + \tfrac{8}{3}rb]\,\mu^2 + \tfrac{8}{3}r[a + b - \tfrac{2}{3}ab]\,\mu -$$

$$[(1 - \tfrac{1}{3}a) - \tfrac{8}{3}ra]\,[(1 - \tfrac{1}{3}b) - \tfrac{8}{3}rb] = 0, \qquad (123)$$

where

$$a = \sin^2\frac{\beta h}{2}$$

and

$$b = \sin^2\frac{\gamma h}{2}.$$

The condition for stability is $|\mu_i| \leqslant 1$ $(i = 1, 2)$. Returning to equation (123), we now write it in the form

$$[(1 - \tfrac{1}{3}a)(1 - \tfrac{1}{3}b) + \tfrac{64}{9}r^2ab]\,(\mu^2 - 1) + \tfrac{8}{3}r(a + b - \tfrac{2}{3}ab)(\mu^2 + \mu + 1) = 0. \qquad (124)$$

Since $r > 0$ and $0 \leqslant a, b \leqslant 1$, the coefficients

$$(1 - \tfrac{1}{3}a)(1 - \tfrac{1}{3}b) + \tfrac{64}{9}r^2ab \ (\equiv c_1)$$

and

$$\tfrac{8}{3}r(a + b - \tfrac{2}{3}ab) \ (\equiv c_2)$$

are both positive, and so we now examine

$$c_1(\mu^2 - 1) + c_2(\mu^2 + \mu + 1) = 0 \quad (c_1, c_2 > 0). \qquad (125)$$

If

$$\mu = v + i\theta \quad (v, \theta \text{ real}),$$

equation (125) gives rise to the two equations

$$(v^2 - \theta^2 - 1)c_1 + (v^2 - \theta^2 + v + 1)c_2 = 0, \tag{126a}$$

$$\theta\,[2vc_1 + (1 + 2v)c_2] = 0, \tag{126b}$$

for v and θ. From equation (126b), it follows that

$$\theta = 0, \text{ or } v = -\frac{\frac{1}{2}c_2}{c_1 + c_2}.$$

From equation (126a),

$$v = \frac{-\frac{1}{2}c_2 \pm (c_1^2 - \frac{3}{4}c_2^2)^{1/2}}{c_1 + c_2} \text{ when } \theta = 0 \quad \left(c_1 > \frac{3}{\sqrt{2}}\,c_2\right)$$

and

$$v^2 + \theta^2 = \frac{c_2 - c_1}{c_1 + c_2} \text{ when } v = -\frac{\frac{1}{2}c_2}{c_1 + c_2} \quad (c_1 < c_2).$$

In both cases,

$$|\mu| = (v^2 + \theta^2)^{1/2} < 1,$$

and so formula (122) is unconditionally stable.

Returning to formula (121), a typical splitting is

$$[1 + (\tfrac{1}{12} - \tfrac{2}{3}r)\delta_{x_1}^2\,]\,U^{n+1*} = -(\tfrac{1}{12} - \tfrac{2}{3}r)\delta_{x_2}^2\,(\alpha U^n + \beta U^{n-1}) +$$

$$\left.\begin{array}{l}\tfrac{2}{3}r[\delta_{x_1}^2 + \delta_{x_2}^2 + \tfrac{1}{2}\delta_{x_1}^2\,\delta_{x_2}^2\,]\,U^n + [1 + (\tfrac{1}{12} + \tfrac{2}{3}r)(\delta_{x_1}^2 + \delta_{x_2}^2)]\,U^{n-1} \\[2ex] \text{and} \\[2ex] [1 + (\tfrac{1}{12} - \tfrac{2}{3}r)\delta_{x_2}^2\,]\,U^{n+1} = U^{n+1*} + (\tfrac{1}{12} - \tfrac{2}{3}r)\delta_{x_2}^2\,(\alpha U^n + \beta U^{n-1}),\end{array}\right\} \tag{127}$$

where $\alpha + \beta = 1$. The first formula in (127) has an order of accuracy h^2 less than the high accuracy composite formula (121).

Exercise

33. Determine the principal truncation error of the first formula in (127) when $\alpha + \beta = 1$.

Finally, since boundary conditions are essential for the solution of implicit difference methods, it is interesting to examine the iteration procedures described by formulae (118) and (121), when the boundary conditions are independent of the time. The iteration procedures converge if

$$U^{n-1} = U^n = U^{n+1} = U_{l,m}$$

for n sufficiently large, and so formulae (118) and (121) reduce to

$$(\delta^2_{x_1} + \delta^2_{x_2})U_{l,m} = 0$$

and

$$(\delta^2_{x_1} + \delta^2_{x_2} + \tfrac{1}{6}\delta^2_{x_1}\delta^2_{x_2})U_{l,m} = 0$$

respectively. These are the five and nine point difference replacements of Laplace's equation, and so, provided sequences of iteration parameters can be found, formulae (118) and (121) can be used as iterative methods for solving Laplace's equation. Parameter sequences leading to efficient iteration using formulae (118) and (121) can be found in Cannon and Douglas (1964).

2.22 Non-linear equations

One of the most gratifying features of the solution of partial differential equations by difference methods is that many of the methods and proofs based on linear equations with constant coefficients carry over directly to non-linear equations. Thus many of the simplest explicit and implicit two-level methods described earlier in this chapter can be used for non-linear equations. If, for example, we consider the parabolic equation

$$\frac{\partial u}{\partial t} = f\left(x, t, u, \frac{\partial u}{\partial x}, \frac{\partial^2 u}{\partial x^2}\right) \quad \left(\frac{\partial f}{\partial u_{xx}} > 0\right), \tag{128}$$

in the region $(0 \leqslant x \leqslant 1) \times (0 \leqslant t \leqslant T)$ subject to initial and boundary conditions, a satisfactory explicit difference replacement of equation (128) is

$$U_m^{n+1} = U_m^n + kf\left[mh, nk, U_m^n, \frac{1}{2h}(U_{m+1}^n - U_{m-1}^n), \frac{1}{h^2}\delta^2_x U_m^n\right]. \tag{129}$$

This difference approximation is very simple to use, but suffers from the disadvantage that the ratio of the time step to the square of the space increment is strictly limited. The limitation depends on the function f, and we have already seen that $k \leqslant \tfrac{1}{2}h^2$ when $f \equiv \partial^2 u/\partial x^2$. We again remind the reader that *if the first space derivative* is important, backward (or forward) difference replacements of $\partial u/\partial x$ should be used in preference to the central difference replacement in (129). Although of lower accuracy, it eliminates the oscillations which may be present when a central difference is used.

The stability limitation can be removed by using an implicit difference method of the Crank–Nicolson type. With reference to equation (128), this is

$$\frac{1}{k} (U_m^{n+1} - U_m^n) = f\left[mh, (n + \tfrac{1}{2})k, \tfrac{1}{2}(U_m^{n+1} + U_m^n), \right.$$

$$\left. \tfrac{1}{2}\left\{ \frac{1}{2h} (U_{m+1}^{n+1} - U_{m-1}^{n+1}) + \frac{1}{2h} (U_{m+1}^n - U_{m-1}^n) \right\}, \frac{1}{2h^2} \delta_x^2(U_m^{n+1} + U_m^n) \right]. \tag{130}$$

Unfortunately, depending on the form of f, the algebraic problem of determining U_m^{n+1} from equation (130) may become quite complicated, since, unless f is very simple, e.g. $f \equiv \partial^2 u/\partial x^2$, the algebraic equations at each time step are non-linear.

Non-linear algebraic equations can be avoided, however, if *two-step* methods of solution are used. For example, if we consider (128) to be Burgers' equation

$$\frac{\partial u}{\partial t} = \epsilon \, \frac{\partial^2 u}{\partial x^2} - u \, \frac{\partial u}{\partial x} \tag{131}$$

where ϵ is a positive constant, the implicit method (130) becomes

$$\frac{1}{k} (U_m^{n+1} - U_m^n) = \frac{\epsilon}{2h^2}\delta_x^2 (U_m^{n+1} + U_m^n) - \tfrac{1}{2}(U_m^{n+1} + U_m^n)\tfrac{1}{2}[\frac{1}{2h} (U_{m+1}^{n+1} - U_{m-1}^{n+1}) +$$

$$\frac{1}{2h} (U_{m+1}^n - U_{m-1}^n)]. \tag{132}$$

This can be rewritten as a two step method of the form

$$\frac{2}{k} (U_m^{n+1/2} - U_m^n) = \frac{\epsilon}{2h^2} \delta_x^2 (U_m^{n+1/2} + U_m^n) - \frac{1}{2h} U_m^n(U_{m+1}^n - U_{m-1}^n), \tag{133}$$

$$\frac{2}{k} (U_m^{n+1} - U_m^{n+1/2}) = \frac{\epsilon}{2h^2}\delta_x^2(U_m^{n+1} + U_m^{n+1/2}) - \frac{1}{2h} U_m^{n+1/2} (U_{m+1}^{n+1/2} - U_{m-1}^{n+1/2}),$$

where the equations to be solved in the two steps of (133) are *linear*. This idea of introducing an intermediate step was first investigated by Douglas and Jones (1963) and further material on methods for solving non-linear algebraic equations can be found in Isaacson and Keller (1967) (chapter 3) and Ortega and Rheinboldt (1970).

It would thus appear that difference methods based on two time levels have poor stability properties if explicit, and are difficult to use computationally if implicit, when applied to the solution of non-linear parabolic equations. Accordingly, we turn to three-level methods in the hope that non-linear equations can be dealt with in a less complicated fashion than the Crank–Nicolson type method given by equation (130). To illustrate the use of three time level schemes, we consider the non-linear parabolic equation

$$b(u) \frac{\partial u}{\partial t} = \frac{\partial}{\partial x}\left(a(u) \frac{\partial u}{\partial x} \right) \quad (a(u) > 0, b(u) > 0). \tag{134}$$

Following equation (113), the simplest difference approximation to equation (134) is

$$b(U_m^n) \frac{1}{2k} (U_m^{n+1} - U_m^{n-1}) = \frac{1}{h^2} \delta_x(a(U_m^n)\delta_x)U_m^n, \tag{135}$$

where the difference operator on the right-hand side is the usual central difference approximation (see (22)). There does not seem much point in considering equation (135) further, because we have already seen that for $a(u) = b(u) = 1$, the difference method is unconditionally unstable.

However, in the linear constant coefficient case, unconditional stability is obtained by replacing $\delta_x^2 U_m^n$ in equation (113) by $\frac{1}{3}\delta_x^2(U_m^{n+1} + U_m^n + U_m^{n-1})$, leading to the implicit scheme (116). Following this procedure, equation (135) is rewritten as

$$b(U_m^n)(U_m^{n+1} - U_m^{n-1}) =$$

$$2r[a(U_{m+1/2}^n)(U_{m+1}^n - U_m^n) - a(U_{m-1/2}^n)(U_m^n - U_{m-1}^n)],$$

and then U_{m+1}^n, U_m^n and U_{m-1}^n are replaced by

$$\tfrac{1}{3}(U_{m+1}^{n+1} + U_{m+1}^n + U_{m+1}^{n-1}),$$

$$\tfrac{1}{3}(U_m^{n+1} + U_m^n + U_m^{n-1})$$

and

$$\tfrac{1}{3}(U_{m-1}^{n+1} + U_{m-1}^n + U_{m-1}^{n-1})$$

respectively, leading to the formula

$$b(U_m^n)(U_m^{n+1} - U_m^{n-1}) =$$

$$\tfrac{2}{3}r[\alpha^+\{(U_{m+1}^{n+1} - U_m^{n+1}) + (U_{m+1}^n - U_m^n) + (U_{m+1}^{n-1} - U_m^{n-1})\} -$$

$$\alpha^-\{(U_m^{n+1} - U_{m-1}^{n+1}) + (U_m^n - U_{m-1}^n) + (U_m^{n-1} - U_{m-1}^{n-1})\}] \tag{136}$$

where

$$\alpha^+ = a\left(\frac{U^n_{m+1} + U^n_m}{2}\right)$$

and

$$\alpha^- = a\left(\frac{U^n_m + U^n_{m-1}}{2}\right).$$

We have replaced

$$a(U^n_{m+1/2}) \text{ and } a(U^n_{m-1/2})$$

by

$$a\left(\frac{U^n_{m+1} + U^n_m}{2}\right) \text{ and } a\left(\frac{U^n_m + U^n_{m-1}}{2}\right)$$

respectively, because the latter expressions involve values of U at grid points. The order of accuracy of the formula is maintained. Formula (136) was studied extensively by Lees (1966), who proved the convergence result for (136) that there exists a constant A, independent of h, k and u, such that

$$\max_{m,n} |U^n_m - u(mh, nk)| \leqslant A(h^2 + k^2),$$

for all sufficiently small h and k where $u(mh, nk)$ is the solution of the differential equation at the grid point (mh, nk). A noteworthy feature of formula (136) is that the system of equations to be solved at $t = (n + 1)k$ is *linear*, and so the complication of solving a set of non-linear equations at each time step, implicit in the Crank–Nicolson type method (130), has been avoided. On the debit side, however, in common with all three-level schemes for parabolic equations extra data is required at $t = k$ before the calculation using (136) can commence.

Exercise

34. Using Taylor expansions about the grid point $x = mh$, $t = nk$, determine the local accuracy of formula (136).

Example 13 Modify equation (135) in order to obtain an explicit formula corresponding to the Du Fort–Frankel formula (115).

From equation (135)

$$b(U^n_m)(U^{n+1}_m - U^{n-1}_m) = 2r[\alpha^+(U^n_{m+1} - U^n_m) - \alpha^-(U^n_m - U^n_{m-1})]$$

$$= 2r[\alpha^+ U^n_{m+1} - (\alpha^+ + \alpha^-)U^n_m + \alpha^- U^n_{m-1}].$$

Du Fort and Frankel proposed that U_m^n be replaced by $\frac{1}{2}(U_m^{n+1} + U_m^{n-1})$, and so the formula becomes

$$b(U_m^n)(U_m^{n+1} - U_m^{n-1}) =$$
$$2r(\alpha^+ U_{m+1}^n + \alpha^- U_{m-1}^n) - r(\alpha^+ + \alpha^-)(U_m^{n+1} + U_m^{n-1}),$$

leading to

$$[b(U_m^n) + r(\alpha^+ + \alpha^-)]\, U_m^{n+1} =$$
$$2r(\alpha^+ U_{m+1}^n + \alpha^- U_{m-1}^n) + [b(U_m^n) - r(\alpha^+ + \alpha^-)]\, U_m^{n-1}, \qquad (137)$$

which is the required result.

The extension of the results on non-linear equations to *two space variables* looks fairly straightforward, at least as far as setting up the formulae is concerned. For example, equation (134) becomes

$$b(u)\,\frac{\partial u}{\partial t} = \frac{\partial}{\partial x_1}\left(a_1(u)\,\frac{\partial u}{\partial x_1}\right) + \frac{\partial}{\partial x_2}\left(a_2(u)\,\frac{\partial u}{\partial x_2}\right)$$

$$(a_1(u) > 0,\, a_2(u) > 0,\, b(u) > 0), \qquad (138)$$

and the difference formula (136) which can be written in the form

$$\left[1 - \frac{2r}{3b}\delta_x(a\delta_x)\right]U_m^{n+1} = \frac{2r}{3b}\delta_x(a\delta_x)U_m^n + \left[1 + \frac{2r}{3b}\delta_x(a\delta_x)\right]U_m^{n-1}$$

extends to two space dimensions to give

$$\left[1 - \frac{2r}{3b}\{\delta_{x_1}(a_1\delta_{x_1}) + \delta_{x_2}(a_2\delta_{x_2})\}\right]U_{l,m}^{n+1}$$

$$= \frac{2r}{3b}\{\delta_{x_1}(a_1\delta_{x_1}) + \delta_{x_2}(a_2\delta_{x_2})\}U_{l,m}^n$$

$$+ \left[1 + \frac{2r}{3b}\{\delta_{x_1}(a_1\delta_{x_1}) + \delta_{x_2}(a_2\delta_{x_2})\}\right]U_{l,m}^{n-1}, \qquad (139)$$

where $b \equiv b(U_{l,m}^n)$ and $a_i \equiv a_i(U_{l,m}^n)$, for $i = 1, 2$. When $a_1 = a_2 = b = 1$, equation (139) reduces to formula (117), and A.D.I. methods of solution of (139) can be set up similarly to those devised for (117).

Finally, although it may appear in general that three-level difference methods have an advantage over two-level schemes in the solution of parabolic equations, it is possible that the introduction of an extra level may cause trouble in particular problems. Each non-linear problem has its own particular difficulties, and the choice between a two-level and a three-level method in a particular case is a difficult one. It is the authors' experience that as many methods as possible, explicit and implicit, two level and three level, should be tried on any non-linear problem, and a critical analysis made of the results obtained.

Chapter 3

Elliptic Equations

3.1 Elliptic equations in two dimensions

Suppose that R is a bounded region in the (x_1, x_2) plane with boundary ∂R. The equation

$$a(x_1, x_2) \frac{\partial^2 u}{\partial x_1^2} + 2b(x_1, x_2) \frac{\partial^2 u}{\partial x_1 \partial x_2} + c(x_1, x_2) \frac{\partial^2 u}{\partial x_2^2} = d\left(x_1, x_2, u, \frac{\partial u}{\partial x_1}, \frac{\partial u}{\partial x_2} \right)$$

(1)

is said to be *elliptic* in R if $b^2 - ac < 0$ for all points (x_1, x_2) in R. Three distinct problems involving equation (1) arise depending on the boundary conditions prescribed on ∂R:

(i) the first boundary value problem, or *Dirichlet problem*, requires a solution u of equation (1) which takes on prescribed values

$$u = f(x_1, x_2)$$

(2)

on the boundary ∂R.

(ii) the second boundary value problem, or *Neumann problem*, where

$$\frac{\partial u}{\partial n} = g(x_1, x_2)$$

(3)

on ∂R. Here $\partial/\partial n$ refers to differentiation along the normal to ∂R directed away from the interior of R.

(iii) the third boundary value problem, or *Robbins problem*, with

$$\alpha(x_1, x_2)u + \beta(x_1, x_2) \frac{\partial u}{\partial n} = \gamma(x_1, x_2)$$

on ∂R, where $\alpha(x_1, x_2) > 0, \beta(x_1, x_2) > 0$ for $(x_1, x_2) \in \partial R$.

102

Before developing finite difference methods of solving elliptic equations, a most useful analytical tool in the study of elliptic partial differential equations will be introduced. This is the *maximum principle* which will be stated for the linear elliptic equation

$$a\frac{\partial^2 u}{\partial x_1^2} + 2b\frac{\partial^2 u}{\partial x_1 \partial x_2} + c\frac{\partial^2 u}{\partial x_2^2} + d\frac{\partial u}{\partial x_1} + e\frac{\partial u}{\partial x_2} = 0,$$

where a, b, c, d and e are functions of the independent variables x_1, x_2. It is clear that in this case any constant represents a solution of the equation. The maximum principle states that the constants are the only solutions which can assume a maximum or minimum value in the interior of the bounded region R. Alternatively, it states that every solution of the elliptic equation achieves its maximum and minimum values on the boundary of R. A full discussion of the maximum principle with respect to elliptic equations can be found in Garabedian (1964).

3.2 Laplace's equation in a square

In order to illustrate how difference methods are used to solve elliptic equations, we consider the simplest non-trivial form of equation (1), viz. Laplace's equation

$$\frac{\partial^2 u}{\partial x_1^2} + \frac{\partial^2 u}{\partial x_2^2} = 0, \qquad (4)$$

subject to the *Dirichlet* condition $u = f(x_1, x_2)$ on the boundary of the unit square $0 \leqslant x_1, x_2 \leqslant 1$. The square region is covered by a grid with sides parallel to the coordinate axes and the grid spacing is h. If $Mh = 1$, the number of internal grid points or nodes is $(M - 1)^2$.

The coordinates of a typical internal grid point are $X_1 = lh$, $X_2 = mh$ (l and m integers), and the value of u at this grid point is denoted by $u_{l,m}$. Using Taylor's theorem, we obtain

$$u_{l+1,m} = \left(u + h\frac{\partial u}{\partial x_1} + \tfrac{1}{2}h^2\frac{\partial^2 u}{\partial x_1^2} + \tfrac{1}{6}h^3\frac{\partial^3 u}{\partial x_1^3} + \tfrac{1}{24}h^4\frac{\partial^4 u}{\partial x_1^4} + \ldots \right)_{l,m}$$

and

$$u_{l-1,m} = \left(u - h\frac{\partial u}{\partial x_1} + \tfrac{1}{2}h^2\frac{\partial^2 u}{\partial x_1^2} - \tfrac{1}{6}h^3\frac{\partial^3 u}{\partial x_1^3} + \tfrac{1}{24}h^4\frac{\partial^4 u}{\partial x_1^4} - \ldots \right)_{l,m}$$

and after addition,

$$u_{l+1,m} + u_{l-1,m} - 2u_{l,m} = \left(h^2\frac{\partial^2 u}{\partial x_1^2} + \tfrac{1}{12}h^4\frac{\partial^4 u}{\partial x_1^4} + \ldots \right)_{l,m}.$$

Similarly,

$$u_{l,m+1} + u_{l,m-1} - 2u_{l,m} = \left(h^2 \frac{\partial^2 u}{\partial x_2^2} + \,^1/_{12}\, h^4 \frac{\partial^4 u}{\partial x_2^4} + \dots \right)_{l,m},$$

and so

$$(u_{l+1,m} + u_{l-1,m} + u_{l,m+1} + u_{l,m-1} - 4u_{l,m})/h^2 =$$

$$\left(\frac{\partial^2 u}{\partial x_1^2} + \frac{\partial^2 u}{\partial x_2^2} \right)_{l,m} + \,^1/_{12}\, h^2 \left(\frac{\partial^4 u}{\partial x_1^4} + \frac{\partial^4 u}{\partial x_2^4} \right)_{l,m} + \dots \quad (5)$$

leading to the *five point* finite difference replacement

$$U_{l+1,m} + U_{l-1,m} + U_{l,m+1} + U_{l,m-1} - 4U_{l,m} = 0,$$

for Laplace's equation, with a local truncation error

$$^1/_{12}\, h^2 \left(\frac{\partial^4 u}{\partial x_1^4} + \frac{\partial^4 u}{\partial x_2^4} \right)_{l,m} + \dots .$$

where $U_{l,m}$ denotes the function satisfying the difference equation at the grid point $X_1 = lh$, $X_2 = mh$. The principal part of the truncation error is meaningful only if $u \in C^{4,4}$ (derivatives of u continuous up to order four in x_1 and x_2). Under these conditions we say that the local truncation error is $0(h^2)$.

Defining the vector **U** to be

$$[U_{1,1} \dots U_{M-1,1}; U_{1,2} \dots U_{M-1,2}; \dots; U_{1,M-1} \dots U_{M-1,M-1}]^T$$

where $[\ ]^T$ denotes the transpose, imposes an ordering on the $(M-1)^2$ unknown grid values. The particular ordering adopted here is the row by row or *natural ordering* of grid points and will be adhered to throughout this chapter. With this ordering, the totality of equations at the $(M-1)^2$ internal nodes of the unit square leads to the matrix equation

$$A\mathbf{U} = \mathbf{k}, \quad (6)$$

where A is a matrix of order $(M-1)^2$ given by

$$A = \begin{bmatrix} B & -J & & & & \mathbf{O} \\ -J & B & -J & & & \\ & \cdot & \cdot & \cdot & & \\ & & \cdot & \cdot & \cdot & \\ & & & \cdot & \cdot & \cdot \\ \mathbf{O} & & & -J & B & -J \\ & & & & -J & B \end{bmatrix}, \quad (7)$$

with J the unit matrix of order $(M-1)$ and B a matrix of order $(M-1)$ given by

$$
B = \begin{bmatrix}
4 & -1 & & & & \\
-1 & 4 & -1 & & & \mathbf{O} \\
& \cdot & \cdot & \cdot & & \\
& & \cdot & \cdot & \cdot & \\
& & & \cdot & \cdot & \cdot \\
& \mathbf{O} & & -1 & 4 & -1 \\
& & & & -1 & 4
\end{bmatrix}.
\tag{8}
$$

The elements of the vector $\mathbf{k}$ depend on the boundary values of $f(x_1, x_2)$ at grid points on the perimeter of the unit square.

The discussion of appropriate techniques for solving equations of the form (6) is deferred until section 3.11. In the intervening sections an account is given of finite difference approximations of more general elliptic equations, the necessary modifications to cope with non-rectangular domains and the implementation of other types of boundary conditions.

When the differential equation (4) has to be solved over a rectangular rather than a square region, it is often appropriate (and in some cases, necessary) to use a rectangular grid with spacings h_1, h_2 in the x_1, x_2 directions respectively.

Exercise

1. Derive the following five-point difference replacement of Laplace's equation at the grid point (lh_1, mh_2)

$$
\theta^{-1}(U_{l+1,m} + U_{l-1,m}) + \theta(U_{l,m+1} + U_{l,m-1}) - 2(\theta + \theta^{-1})U_{l,m} = 0
$$

where h_1, h_2 are the grid spacings in the x_1, x_2 directions respectively and $\theta = h_1/h_2$ is the mesh ratio.

Show that the local truncation error of this formula is

$$
\tfrac{1}{12}\left(h_1^2 \frac{\partial^4 u}{\partial x_1^4} + h_2^2 \frac{\partial^4 u}{\partial x_2^4} \right) + \dots \ .
$$

3.3 The Neumann Problem

We now consider the Neumann boundary value problem consisting of Laplace's equation

$$
\frac{\partial^2 u}{\partial x_1^2} + \frac{\partial^2 u}{\partial x_2^2} = 0
\tag{9}
$$

in the unit square $0 \leqslant x_1, x_2 \leqslant 1$ subject to the boundary condition

$$
\frac{\partial u}{\partial n} = g(x_1, x_2)
\tag{10}
$$

on the boundary of the square, where $\partial/\partial n$ denotes differentiation along the normal to the boundary directed away from the interior of the square. The normal is not well-defined at the vertices and indeed $g(x_1, x_2)$ will not be continuous there. In the latter case we take the value of $g(x_1, x_2)$ to be the average of the discontinuous values.

A solution to the problem defined by (9) and (10) will exist only when

$$\int_L g(x_1, x_2)\,dl = 0 \qquad (11)$$

where L is the perimeter of the square and dl an element of length along L. The solution will be unique only up to an additive constant, for it is easily seen that if $u(x_1, x_2)$ is any solution of (9) and (10) then so too is $u(x_1, x_2) + c$ for any value of the constant c. Uniqueness therefore requires that u be prescribed at some arbitrary point in the domain.

A uniform grid of spacing h is superimposed on the region with the grid lines parallel to the sides of the square, and if $Mh = 1$, the number of grid points at which u has to be determined is $(M + 1)^2$. The coordinates of a typical grid point are $X_1 = lh$, $X_2 = mh$, and the value of u at this grid point satisfying the difference equation is $U_{l,m}\,(0 \leqslant l,m \leqslant M)$.

Laplace's equation is replaced at internal grid points by

$$(\delta^2_{x_1} + \delta^2_{x_2})U_{l,m} = 0, \qquad (12)$$

and the derivative boundary condition at grid points on the boundary $x_1 = 0$ by

$$\frac{1}{2h}(U_{-1,m} - U_{1,m}) = g_{0,m}, \quad (m = 0, 1, 2, \ldots, M), \qquad (13)$$

where $g_{0,m} \equiv g(0, mh)$ and similar expressions for condition (10) on the boundaries $x_1 = 1$ and $x_2 = 0,1$. The value $U_{-1,m}$ which lies outside the square is eliminated between equations (12) and (13) applied at the corresponding grid point on $x_1 = 0$. In this manner, the boundary conditions are incorporated into the difference scheme. The totality of equations at the $(M + 1)^2$ grid points of the square leads to the matrix equation

$$A\mathbf{U} = 2h\mathbf{G}, \qquad (14)$$

where A is a matrix of order $(M + 1)^2$ given by

$$A = \begin{bmatrix} B & -2I & & & & \\ -I & B & -I & & & \\ & & \ddots & \ddots & \ddots & \\ & & & \ddots & \ddots & \ddots \\ & & & -I & B & -I \\ & & & & -2I & B \end{bmatrix}$$

with I the unit matrix of order $(M + 1)$, and B a matrix of order $(M + 1)$ given by

$$B = \begin{bmatrix} 4 & -2 & & & & \\ -1 & 4 & -1 & & & \\ & & \ddots & \ddots & \ddots & \\ & & & \ddots & \ddots & \ddots \\ & & & -1 & 4 & -1 \\ & & & & -2 & 4 \end{bmatrix}.$$

The vectors $\mathbf{U}$ and $\mathbf{G}$ of equation (14) are given by

$$[U_{0,0} \ldots U_{M,0} ; U_{0,1} \ldots U_{M,1} ; \ldots ; U_{0,M} \ldots U_{M,M}]^T$$

and

$$[2g_{0,0}, g_{1,0} \ldots g_{M-1,0}, 2g_{M,0} ; g_{0,1}, 0 \ldots 0, g_{M,1} ; \ldots ;$$
$$g_{0,M-1}, 0 \ldots 0, g_{M,M-1} ; 2g_{0,M}, g_{1,M} \ldots g_{M-1,M}, 2g_{M,M}]^T$$

respectively, where $[\ ]^T$ denotes the transpose of the vector, and $g_{l,m} \equiv g(lh, mh)$ with $0 \leqslant l, m \leqslant M$.

In contrast to the Dirichlet problem, the matrix A of (14) is singular. This can best be demonstrated by showing that $A\mathbf{V} = 0$ where $\mathbf{V}$ is the vector each of whose components is unity. Hence, for a solution $\mathbf{U}$ of (14) to exist, $\mathbf{G}$ must be a linear combination of columns of A in which case $\mathbf{U} + c\mathbf{V}$ will also be a solution for any scalar constant c. This corresponds precisely to the situation outlined earlier for the differential equation. An example will now be given to illustrate the foregoing theory.

Example 1 Solve the Neumann problem for Laplace's equation in the unit square with

$$h = \tfrac{1}{2}.$$

108

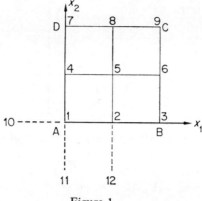

Figure 1

In this case, using the notation in figure 1, equation (14) gives

$$
\begin{bmatrix}
4 & -2 & 0 & -2 & 0 & 0 & 0 & 0 & 0 \\
-1 & 4 & -1 & 0 & -2 & 0 & 0 & 0 & 0 \\
0 & -2 & 4 & 0 & 0 & -2 & 0 & 0 & 0 \\
-1 & 0 & 0 & 4 & -2 & 0 & -1 & 0 & 0 \\
0 & -1 & 0 & -1 & 4 & -1 & 0 & -1 & 0 \\
0 & 0 & -1 & 0 & -2 & 4 & 0 & 0 & -1 \\
0 & 0 & 0 & -2 & 0 & 0 & 4 & -2 & 0 \\
0 & 0 & 0 & 0 & -2 & 0 & -1 & 4 & -1 \\
0 & 0 & 0 & 0 & 0 & -2 & 0 & -2 & 4
\end{bmatrix}
\begin{bmatrix}
U_1 \\ U_2 \\ U_3 \\ U_4 \\ U_5 \\ U_6 \\ U_7 \\ U_8 \\ U_9
\end{bmatrix}
= 2h
\begin{bmatrix}
2g_1 \\ g_2 \\ 2g_3 \\ g_4 \\ 0 \\ g_6 \\ 2g_7 \\ g_8 \\ 2g_9
\end{bmatrix}
\tag{15}
$$

First note that when these nine scalar equations are multiplied by

¼, ½, ¼, ½, 1, ½, ¼, ½, ¼

respectively, the resulting coefficient matrix is symmetric. When the nine scaled equations are now added together, we obtain

$$
\sum_{i=1}^{9} g_i = 0 \qquad (g_5 \equiv 0) \tag{16}
$$

verifying the fact that A is singular. A solution of (15) will exist only if condition (16) is satisfied, otherwise the equations are inconsistent.

It can be shown by approximating the integral in (11) using the repeated trapezoidal rule that

$$
\int_L g(x_1, x_2)\,\mathrm{d}l \approx h \sum_{i=1}^{9} g_i .
$$

Thus, even when (16) is not exactly satisfied, $\sum g_i$ should be small. It is therefore reasonable to set one component, say U_9, equal to an arbitrary constant and solve the first eight of equations (15) for $U_1, U_2, \ldots, U_8$.

3.4 Mixed Boundary Conditions

In the most commonly occurring problems involving Laplace's equation on a region R, the boundary is divided into a finite number of arcs and a different type of boundary condition is imposed on each arc. This situation may be illustrated by solving Laplace's equation

$$\frac{\partial^2 u}{\partial x_1^2} + \frac{\partial^2 u}{\partial x_2^2} = 0$$

in the region $R = [\![0 \leqslant x_1, x_2 \leqslant 1]\!]$, subject to the *Robbins* type boundary conditions

$$\frac{\partial u}{\partial x_1} - p(x_2)u = f_0(x_2), \quad x_1 = 0 \quad (0 \leqslant x_2 \leqslant 1),$$

$$\frac{\partial u}{\partial x_2} - q(x_1)u = f_1(x_1), \quad x_2 = 0 \qquad (0 \leqslant x_1 \leqslant 1),$$

and the *Dirichlet* type boundary conditions

$$u = g(x_1, x_2), \quad \left\{ \begin{array}{l} x_1 = 1 \quad (0 \leqslant x_2 \leqslant 1) \\[2mm] x_2 = 1 \quad (0 \leqslant x_1 \leqslant 1) \end{array} \right.,$$

where p, q, f_0, f_1 and g are prescribed functions.

As usual, the square region R is covered by a rectilinear grid, with grid points $X_1 = lh$, $X_2 = mh$ where $l, m = 0, 1, 2, \ldots, M$ and $Mh = 1$. The Laplacian operator is again replaced by the five-point difference formula

$$(\delta_{x_1}^2 + \delta_{x_2}^2)U_{l,m} = 0 \tag{17}$$

where $U_{l,m}$ is the approximation to the function u at the grid point (lh, mh). Equation (17) holds in its original form at the internal grid points and in a form modified by difference approximations to the boundary conditions at grid points on the boundaries $x_1 = 0$ and $x_2 = 0$.

For a typical grid point $(0, mh)$ on the boundary $x_1 = 0$, application of (17) along with the discrete approximation to the boundary condition

$$\frac{\partial u}{\partial x_1} - p(x_2)u = f_0(x_2)$$

leads to

$$U_{-1,m} + U_{1,m} + U_{0,m+1} + U_{0,m-1} - 4U_{0,m} = 0$$

and

$$(U_{1,m} - U_{-1,m})/2h - p_m U_{0,m} = f_{0,m}$$

respectively, where $p_m \equiv p(mh)$ and $f_{0,m} \equiv f_0(mh)$. Eliminating $U_{-1,m}$ between these two equations gives

$$2U_{1,m} + U_{0,m+1} + U_{0,m-1} - (4 + 2hp_m)U_{0,m} = 2hf_{0,m} \qquad m = 1, 2, \ldots, M-1$$

which no longer involves grid values outside R.

A similar approach can be adopted for the boundary $x_2 = 0$ and the only exceptional point is the origin where two boundaries meet. At $l = m = 0$ we have

$$U_{-1,0} + U_{1,0} + U_{0,1} + U_{0,-1} - 4U_{0,0} = 0,$$

$$U_{1,0} - U_{-1,0} - 2hp_0 U_{0,0} = 2hf_{0,0},$$

and

$$U_{0,1} - U_{0,-1} - 2hq_0 U_{0,0} = 2hf_{1,0},$$

from which we can eliminate both $U_{0,-1}$ and $U_{-1,0}$ to give

$$2U_{1,0} + 2U_{0,1} - (4 + 2hp_0 + 2hq_0)U_{0,0} = 2h(f_{0,0} + f_{1,0}).$$

When assembling the difference equations into matrix form it is convenient to multiply the equations which hold for $l = 0$ and $m = 0$ by the factor ½ and the equation at $l = m = 0$ by ¼. This renders the coefficient matrix symmetric and, when the grid values are ordered along horizontal lines, there results

$$A\mathbf{U} = h\mathbf{G}$$

where A is the matrix of order M^2 given by

$$A = \begin{bmatrix} E_0 & K & & & & \\ K & E_1 & K & & & \\ & K & . & . & & \\ & & . & . & . & \\ & & & . & . & K \\ & & & & K & E_{M-1} \end{bmatrix}$$

and $E_0, E_1, \ldots, E_{M-1}, K$ are the matrices of order M defined by

$$E_0 = \begin{bmatrix} -[1 + \tfrac{1}{2}h(p_0 + q_0)] & \tfrac{1}{2} & & & & & \bigcirc \\ \tfrac{1}{2} & -[2 + hq_1] & \tfrac{1}{2} & & & & \\ & \tfrac{1}{2} & & & & & \\ & & & \cdot & \cdot & \cdot & \tfrac{1}{2} \\ \bigcirc & & & & \tfrac{1}{2} & -[2 + hq_{M-1}] \end{bmatrix},$$

$$E_m = \begin{bmatrix} -[2 + hp_m] & 1 & & & \bigcirc \\ 1 & -4 & 1 & & \\ & 1 & \cdot & & \\ & & \cdot & \cdot & 1 \\ \bigcirc & & & 1 & -4 \end{bmatrix} \quad (m = 1, 2, \ldots, M-1),$$

and

$$K = \begin{bmatrix} \tfrac{1}{2} & & & \bigcirc \\ & 1 & & \\ & & 1 & \\ & & & \cdot \\ \bigcirc & & & & 1 \end{bmatrix}.$$

The vector **U** is defined by

$$\mathbf{U} = \left[U_{0,0},\ U_{1,0} \ldots U_{M-1,0};\ U_{0,1}, \ldots, U_{M-1,1}; \ldots; U_{M-1,0}, \ldots, U_{M-1,M-1} \right]^T$$

and **G** depends on the functions f_0, f_1 and g evaluated at boundary grid points.

3.5 Non-Rectangular Regions

When the region R is rectangular (or a union of rectangles) with sides parallel to grid lines, the standard difference replacements of an elliptic problem can be applied at all internal grid points. When R is non-rectangular, internal grid points adjacent to the boundary require special treatment. Such a grid point is depicted in figure 2.

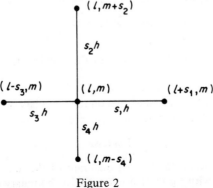

Figure 2

Taylor expansions about the point (lh, mh) give

$$u_{l+s_1,m} = [u + s_1 h \frac{\partial u}{\partial x_1} + \tfrac{1}{2} s_1^2 h^2 \frac{\partial^2 u}{\partial x_1^2} + 0(h^3)]_{l,m} \tag{18}$$

$$u_{l,m+s_2} = [u + s_2 h \frac{\partial u}{\partial x_2} + \tfrac{1}{2} s_2^2 h^2 \frac{\partial^2 u}{\partial x_2^2} + 0(h^3)]_{l,m} \tag{19}$$

$$u_{l-s_3,m} = [u - s_3 h \frac{\partial u}{\partial x_1} + \tfrac{1}{2} s_3^2 h^2 \frac{\partial^2 u}{\partial x_1^2} + 0(h^3)]_{l,m} \tag{20}$$

and

$$u_{l,m-s_4} = [u - s_4 h \frac{\partial u}{\partial x_2} + \tfrac{1}{2} s_4^2 h^2 \frac{\partial^2 u}{\partial x_2^2} + 0(h^3)]_{l,m} \tag{21}$$

To obtain a difference approximation for Laplace's equation, $\partial u/\partial x_1$ and $\partial u/\partial x_2$ are eliminated between (18), (20) and (19), (21) respectively to give

$$\frac{\partial^2 u}{\partial x_1^2} = \frac{2}{h^2} \{s_3(u_{l+s_1,m} - u_{l,m}) + s_1(u_{l-s_3,m} - u_{l,m})\}/s_1 s_3(s_1 + s_3) + 0(h)$$

and

$$\frac{\partial^2 u}{\partial x_2^2} = \frac{2}{h^2} \{s_4(u_{l,m+s_2} - u_{l,m}) + s_2(u_{l,m-s_4} - u_{l,m})\}/s_2 s_4(s_2 + s_4) + 0(h).$$

Thus, Laplace's equation is replaced by the five-point formula

$$\beta_1 U_{l+s_1,m} + \beta_2 U_{l,m+s_2} + \beta_3 U_{l-s_3,m} + \beta_4 U_{l,m-s_4} - \beta_0 U_{l,m} = 0 \tag{22}$$

where

$$\beta_1 = 2/(s_1 + s_3)s_1, \qquad \beta_2 = 2/(s_2 + s_4)s_2$$

$$\beta_3 = 2/(s_1 + s_3)s_3, \qquad \beta_4 = 2/(s_2 + s_4)s_4$$

and $\beta_0 = \sum_{i=1}^{4} \beta_i$. Note that (22) reduces to the familiar five-point formula when $s_1 = s_2 = s_3 = s_4 = 1$.

When *Dirichlet* conditions are specified on the boundary of R, difference approximations of the form (22) constitute the only necessary modifications.

Exercise

2. Solve Laplace's equation in the first quadrant of the unit circle $R = \{x_1 > 0, x_2 > 0, x_1^2 + x_2^2 < 1\}$ with a grid of size $h = \frac{1}{4}$. The boundary conditions are

$$u(0, x_2) = 0 \quad (0 < x_2 < 1),$$

$$\frac{\partial u}{\partial x_2} = 0 \qquad (x_2 = 0, \, 0 < x_1 < 1),$$

and on the circular arc

$$u = 16x_1^5 - 20x_1^3 + 5x_1 \quad (0 < x_1 < 1).$$

[The exact solution to this problem is given by $u = x_1^5 - 10x_1^3 x_2^2 + 5x_1 x_2^4$. The maximum error, taken over all grid points is 0.048 and the computed value at the grid point (½,½) is −0.102 compared to the exact value u(½,½) = −0.125.]

The imposition of *Neumann* or *Robbins* conditions on a boundary which is not parallel to grid lines presents considerable additional difficulty. As an example, consider the Neumann problem in the circular region, illustrated in figure 3. The circle is of unit radius and is covered by a rectilinear grid, with grid spacing $h = {}^2/_3$.

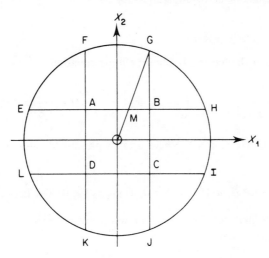

Figure 3

The internal grid points are A, B, C and D, and at B the difference replacement of Laplace's equation is, by equation (22),

$$\frac{2}{1+a}(U_A + U_C) + \frac{2}{a(1+a)}(U_H + U_G) - \frac{4}{a}U_B = 0,$$

where BH = BG = $ah\,[a = (2\sqrt{2} - 1)/3]$ and U_A, U_C, ... are the values of U at the grid points A, C, ... The values of U_G and U_H are now obtained from $(\partial u/\partial n)_G$ and $(\partial u/\partial n)_H$ in the following manner. First

$$\left(\frac{\partial u}{\partial n}\right)_G \approx \frac{1}{MG}(U_G - U_M),$$

but

$$U_{\mathrm{M}} \approx \frac{\mathrm{MB}}{\mathrm{AB}} U_{\mathrm{A}} + \frac{\mathrm{AM}}{\mathrm{AB}} U_{\mathrm{B}} \text{ (linear interpolation)},$$

and so

$$U_{\mathrm{G}} = \frac{\mathrm{MB}}{\mathrm{AB}} U_{\mathrm{A}} + \frac{\mathrm{AM}}{\mathrm{AB}} U_{\mathrm{B}} + \mathrm{MG}\left(\frac{\partial u}{\partial n}\right)_{\mathrm{G}},$$

and similarly for U_{H}. The other internal grid points A, C, and D are dealt with in the same manner as B. For more general regions with curved boundaries, a procedure is adopted for setting up the difference systems similar to that used above for the circle. A more complete account giving other approximations of higher accuracy can be found in Bramble and Hubbard (1965).

3.6 Self-adjoint elliptic equations

We now consider the linear self-adjoint elliptic equation

$$\frac{\partial}{\partial x_1}\left(a_1(x_1, x_2)\frac{\partial u}{\partial x_1} \right) + \frac{\partial}{\partial x_2}\left(a_2(x_1, x_2)\frac{\partial u}{\partial x_2} \right) - f(x_1, x_2)u = g(x_1, x_2) \quad (23)$$

defined for a bounded region R where u takes prescribed values on the boundary ∂R of R, and

$$a_1(x_1, x_2) > 0, \ a_2(x_1, x_2) > 0, \ f(x_1, x_2) \geqslant 0, \text{ for } (x_1, x_2) \in R.$$

For a typical internal grid point, we follow the treatment of section 2.3 and use the difference replacement

$$\frac{1}{h^2} \delta_{x_1}(a_1\delta_{x_1} U_{l,m}) + \frac{1}{h^2} \delta_{x_2}(a_2\delta_{x_2} U_{l,m}) - f_{l,m} U_{l,m} = g_{l,m} \quad (24)$$

which may be written in the form

$$\alpha_1 U_{l+1,m} + \alpha_2 U_{l,m+1} + \alpha_3 U_{l-1,m} + \alpha_4 U_{l,m-1} - \alpha_0 U_{l,m} = h^2 g_{l,m} \quad (25)$$

with

$$\alpha_1 = (a_1)_{l+\frac{1}{2},m}, \ \alpha_2 = (a_2)_{l,m+\frac{1}{2}}, \ \alpha_3 = (a_1)_{l-\frac{1}{2},m}, \ \alpha_4 = (a_2)_{l,m-\frac{1}{2}}$$

and

$$\alpha_0 = \sum_{i=1}^{4} \alpha_i + h^2 f_{l,m}.$$

Exercise

3. Show that the formula (24) has a local truncation error of $0(h^2)$.

Example 2 Solve the problem of Exercise 2 by transforming Laplace's equation into polar co-ordinates (r, θ) where $x_1 = r\cos\theta$ and $x_2 = r\sin\theta$.

In polar co-ordinates Laplace's equation takes the form

$$\frac{1}{r}\frac{\partial}{\partial r}\left(r\frac{\partial u}{\partial r}\right) + \frac{1}{r^2}\frac{\partial^2 u}{\partial \theta^2} = 0.$$

The first quadrant of the unit circle $0 < r < 1, 0 < \theta < \pi/2$ is covered by grid lines $r = l\Delta r, \theta = m\Delta\theta, l,m = 1, 2, \ldots, M - 1$, where $\Delta r = 1/M$ and $\Delta\theta = \pi/2M$. Following the procedure outlined above we obtain the difference formula

$$\alpha^{-1}(r_{l+\frac{1}{2}}U_{l+1,m} + r_{l-\frac{1}{2}}U_{l-1,m}) + \alpha(U_{l,m+1} + U_{l,m-1})/r_l - 2(\alpha + r_l/\alpha)U_{l,m} = 0$$

$$(26)$$

for $l = 1, 2, \ldots, M - 1, m = 0, 1, \ldots, M - 1$ where $r_l \equiv l\Delta r$ and $\alpha = \Delta r/\Delta\theta$. The boundary ∂R is formed by the grid lines $m = 0, M$ and $l = 0, M$ and application of the Dirichlet condition gives

$$U_{0,m} = U_{l,0} = 0 \quad (l,m = 0, 1, \ldots, M),$$

and

$$U_{M,m} = \cos^5 m\Delta\theta - 20\cos^3 m\Delta\theta + 5\cos m\Delta\theta \quad (m = 0, 1, \ldots, M).$$

The Neumann condition on $x_2 = 0$ is transformed into

$$\frac{\partial u}{\partial \theta} = 0 \quad \text{for } \theta = 0 \quad (0 < r < 1),$$

and this is approximated in the usual way to give

$$U_{l,-1} = U_{l,1} \quad (l = 1, 2, \ldots, M - 1).$$

Equation (26) together with these boundary conditions constitutes a system of $M(M - 1)$ equations for the determination of $U_{l,m}$ ($l = 1, 2, \ldots, M - 1$ and $m = 0, 1, \ldots, M - 1$).

It is interesting to note that when these equations are solved with $M = 4$, the maximum error at grid points is 0.067 which is in fact larger than that obtained in Exercise 2 using a Cartesian grid.

For a grid point adjacent to the boundary (see figure 2), we use the difference approximations

$$\left[\frac{\partial}{\partial x_1} \left(a_1(x_1, x_2) \frac{\partial u}{\partial x_1} \right) \right]_{l,m} \approx$$

$$\frac{(a_1)_{l+\frac{1}{2}s_1,m} \left(\dfrac{U_{l+s_1,m} - U_{l,m}}{s_1 h} \right) - (a_1)_{l+\frac{1}{2}s_3,m} \left(\dfrac{U_{l,m} - U_{l-s_3,m}}{s_3 h} \right)}{\frac{1}{2}(s_1 + s_3)h}$$

and

$$\left[\frac{\partial}{\partial x_2} \left(a_2(x_1, x_2) \frac{\partial u}{\partial x_2} \right) \right]_{l,m} \approx$$

$$\frac{(a_2)_{l,m+\frac{1}{2}s_2} \left(\dfrac{U_{l,m+s_2} - U_{l,m}}{s_2 h} \right) - (a_2)_{l,m-\frac{1}{2}s_4} \left(\dfrac{U_{l,m} - U_{l,m-s_4}}{s_4 h} \right)}{\frac{1}{2}(s_2 + s_4)h}$$

where $0 < s_i \leqslant 1$ $(i = 1, 2, 3, 4)$, and so at such a grid point equation (25) becomes

$$\beta_1 U_{l+s_1,m} + \beta_2 U_{l,m+s_2} + \beta_3 U_{l-s_3,m} + \beta_4 U_{l,m-s_4} -$$
$$\left(\sum_{i=1}^{4} \beta_i + h^2 f_{l,m} \right) U_{l,m} = h^2 g_{l,m} , \qquad (27)$$

where

$$\beta_1 = \frac{2}{s_1(s_1 + s_3)} (a_1)_{l+\frac{1}{2}s_1,m}, \quad \beta_2 = \frac{2}{s_2(s_2 + s_4)} (a_2)_{l,m+\frac{1}{2}s_2},$$

$$\beta_3 = \frac{2}{s_3(s_1 + s_3)} (a_1)_{l-\frac{1}{2}s_3,m}, \quad \beta_4 = \frac{2}{s_4(s_2 + s_4)} (a_2)_{l,m-\frac{1}{2}s_4}.$$

The difference equation (27) reduces to (25) when $s_1 = s_2 = s_3 = s_4 = 1$. The totality of equations consisting of equation (26) at internal grid points and equation (27) with appropriate values of s_i $(i = 1, 2, 3, 4)$ at grid points in the vicinity of the boundary leads to the matrix equation,

$$A\mathbf{U} = \mathbf{g},$$

where A is a *diagonally dominant* matrix.

3.7 Alternative methods for constructing difference formulae

Hitherto difference formulae have been constructed by replacing each term in the differential equation by an appropriate difference quotient. Since such a technique is limited to basically uniform grids it is sometimes necessary to employ methods having wider applicability.

Suppose the elliptic equation is

$$Lu(x_1, x_2) \equiv a_1 \frac{\partial^2 u}{\partial x_1^2} + 2b \frac{\partial^2 u}{\partial x_1 \partial x_2} + a_2 \frac{\partial^2 u}{\partial x_2^2} + c_1 \frac{\partial u}{\partial x_1} + c_2 \frac{\partial u}{\partial x_2} + du = 0,$$

$$(x_1, x_2) \in R, \qquad (28)$$

where the coefficients depend on x_1, x_2 and the region R is covered by an arbitrary grid where the distance between neighbouring grid points is not constant. A linear algebraic expression is constructed to represent the differential equation at any grid point P by means of Taylor series. It is of the form

$$L_h U \equiv \alpha_0 U_P - \sum_{i=1}^{s} \alpha_i U_{Q_i} = 0 \qquad (29)$$

where Q_i $(i = 1, 2, \ldots, s)$ are neighbouring grid points to P and the terms U_P, U_{Q_i} represent the values of U at the grid points P, Q_i $(i = 1, 2, \ldots, s)$ respectively. Due to the irregular nature of the grid, the coefficients $\alpha_0, \alpha_1, \ldots, \alpha_s$ will depend on the location of P but the notational dependence is suppressed in the interests of clarity.

For convenience, we choose $P \equiv (0,0)$, $Q_i \equiv (\xi_i h, \eta_i h)$ $(i = 1, 2, \ldots, s)$, and a Taylor expansion of $u(\xi_i h, \eta_i h)$ about P leads to

$$u_{Q_i} \equiv u(\xi_i h, \eta_i h) = u_P + \xi_i h \left(\frac{\partial u}{\partial x_1} \right)_P + \eta_i h \left(\frac{\partial u}{\partial x_2} \right)_P + \tfrac{1}{2}\xi_i^2 h^2 \left(\frac{\partial^2 u}{\partial x_1^2} \right)_P +$$

$$\xi_i \eta_i h^2 \left(\frac{\partial^2 u}{\partial x_1 \partial x_2} \right)_P + \tfrac{1}{2}\eta_i^2 h^2 \left(\frac{\partial^2 u}{\partial x_2^2} \right)_P + \ldots .$$

Substitution of these values of u_{Q_i} $(i = 1, 2, \ldots, s)$ into

$$\alpha_0 u_P - \sum_{i=1}^{s} \alpha_i u_{Q_i}$$

leads to

$$u_P(\Sigma \alpha_i - \alpha_0) + h \left(\frac{\partial u}{\partial x_1} \right)_P \Sigma \xi_i \alpha_i + h \left(\frac{\partial u}{\partial x_2} \right)_P \Sigma \eta_i \alpha_i +$$

$$\tfrac{1}{2}h^2 \left(\frac{\partial^2 u}{\partial x_1^2} \right)_P \Sigma \xi_i^2 \alpha_i + h^2 \left(\frac{\partial^2 u}{\partial x_1 \partial x_2} \right)_P \Sigma \xi_i \eta_i \alpha_i + \tfrac{1}{2}h^2 \left(\frac{\partial^2 u}{\partial x_2^2} \right)_P \Sigma \eta_i^2 \alpha_i + \ldots,$$

$$(30)$$

where all summations are from $i = 1$ to $i = s$. Equating the coefficients of like terms in equations (28) and (30) leads to

E

$$\left.\begin{aligned}
\Sigma\alpha_i - \alpha_0 &= d \\
\Sigma\xi_i\alpha_i &= c_1/h \\
\Sigma\eta_i\alpha_i &= c_2/h \\
\tfrac{1}{2}\Sigma\xi_i^2\alpha_i &= a_1/h^2 \\
\tfrac{1}{2}\Sigma\xi_i\eta_i\alpha_i &= b/h^2 \\
\tfrac{1}{2}\Sigma\eta_i^2\alpha_i &= a_2/h^2
\end{aligned}\right\}
\tag{31}$$

a system of six equations in the unknowns $\alpha_0, \alpha_1, \alpha_2, \ldots, \alpha_5$. It would appear that P requires at least five neighbouring points Q_i ($i = 1, 2, \ldots, 5$) in order to satisfy the system (31). In general, $L_h U$ can be made a better representation of Lu either by introducing more neighbouring points Q_i or by using a rectangular grid with equal spacings.

Example 3. Show in the case of Laplace's equation on a square grid that the four neighbouring points

$$(h, 0), (0, h), (-h, 0), (0, -h)$$

are sufficient to satisfy the system (31).

Here,

$$\begin{aligned}
\xi_1 &= 1, & \eta_1 &= 0 \\
\xi_2 &= 0, & \eta_2 &= 1 \\
\xi_3 &= -1, & \eta_3 &= 0 \\
\xi_4 &= 0, & \eta_4 &= -1.
\end{aligned}$$

Substitution into the system (31) gives

$$\begin{aligned}
\alpha_1 + \alpha_2 + \alpha_3 + \alpha_4 &= \alpha_0 \\
\alpha_1 - \alpha_3 &= 0 \\
\alpha_2 - \alpha_4 &= 0 \\
\alpha_1 + \alpha_3 &= 2/h^2 \\
\alpha_2 + \alpha_4 &= 2/h^2,
\end{aligned}$$

leading to

$$\alpha_0 = 4/h^2, \quad \alpha_1 = \alpha_2 = \alpha_3 = \alpha_4 = 1/h^2.$$

Also, from equation (30) the next four equations in the system (31) are

$$\begin{aligned}
\tfrac{1}{6}\Sigma\xi_i^3\alpha_i &= 0 \\
\tfrac{1}{2}\Sigma\xi_i^2\eta_i\alpha_i &= 0 \\
\tfrac{1}{2}\Sigma\xi_i\eta_i^2\alpha_i &= 0 \\
\tfrac{1}{6}\Sigma\eta_i^3\alpha_i &= 0,
\end{aligned}$$

which are also satisfied. The close agreement between $L_h U$ and Lu in this example is caused by the symmetry of the differential equation and the grid.

Exercise

4. Find the values of α_0, α_1, α_2, α_3, α_4 in the case of Laplace's equation on a rectangular grid for the four neighbouring points $(2h, 0)$, $(0, h)$, $(-2h, 0)$, $(0, -h)$.

A popular alternative derivation of finite difference approximations is provided by the 'box integration' scheme. To simplify the ensuing description we shall assume that the grid lines are aligned parallel to the x_1, x_2 axes (but not necessarily equally spaced) and that the elliptic equation is self-adjoint, so that it may be written in the form

$$Lu = \frac{\partial}{\partial x_1}\left(a_1 \frac{\partial u}{\partial x_1}\right) + \frac{\partial}{\partial x_2}\left(a_2 \frac{\partial u}{\partial x_2}\right) - du = 0 \qquad (32)$$

where the coefficients again depend on x_1, x_2.

Consider now a typical grid point P and its four nearest neighbours Q_i ($i = 1, \ldots, 4$). For convenience we assume that the co-ordinates of P are $(0, 0)$ and those of Q_i ($i = 1, 2, 3, 4$) are $(s_1 h, 0)$, $(0, s_2 h)$, $(-s_3 h, 0)$ and $(0, -s_4 h)$ respectively (see figure 4). Let N_i denote the mid-points of PQ_i ($i = 1, 2, 3, 4$) respectively, then integration of each term of (32) over the shaded region S shown in figure 4 leads to

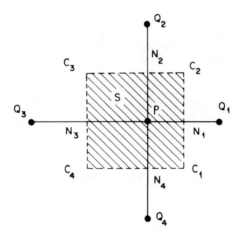

Figure 4

$$\int_S\int \frac{\partial}{\partial x_1}\left(a_1 \frac{\partial u}{\partial x_1}\right) dx_1\, dx_2 =$$

$$\int_{x_2 = -\frac{1}{2}s_4 h}^{\frac{1}{2}s_2 h} \left(a_1 \frac{\partial u}{\partial x_1}\right)\bigg|_{x_1 = \frac{1}{2}s_1 h} dx_2 - \int_{x_2 = -\frac{1}{2}s_4 h}^{\frac{1}{2}s_2 h} \left(a_1 \frac{\partial u}{\partial x_1}\right)\bigg|_{x_1 = -\frac{1}{2}s_3 h} dx_2$$

The integrals on the right are now approximated by the mid-point rule to give

$$\frac{h}{2}(s_2 + s_4)\left\{a_1(N_1)\left.\frac{\partial u}{\partial x_1}\right|_{N_1} - a_1(N_3)\left.\frac{\partial u}{\partial x_1}\right|_{N_3}\right\}, \tag{33}$$

whereupon, using the replacements

$$\left.\frac{\partial u}{\partial x_1}\right|_{N_1} \approx \frac{1}{s_1 h}(U_{Q_1} - U_P) \quad \text{and} \quad \left.\frac{\partial u}{\partial x_1}\right|_{N_3} \approx \frac{1}{s_3 h}(U_P - U_{Q_3})$$

we obtain

$$\int_S\int \frac{\partial}{\partial x_1}\left(a_1 \frac{\partial u}{\partial x_1}\right) dx_1 dx_2 \approx$$

$$\tfrac{1}{2}(s_2 + s_4)\left\{\frac{a_1(N_1)}{s_1}(U_{Q_1} - U_P) - \frac{a_1(N_3)}{s_3}(U_P - U_{Q_3})\right\}. \tag{34}$$

Similarly

$$\int_S\int \frac{\partial}{\partial x_3}\left(a_2 \frac{\partial u}{\partial x_2}\right) dx_1 dx_2 \approx$$

$$\tfrac{1}{2}(s_1 + s_3)\left\{\frac{a_2(N_2)}{s_2}(U_{Q_2} - U_P) - \frac{a_2(N_4)}{s_4}(U_P - U_{Q_4})\right\}. \tag{35}$$

Finally

$$\int_S\int d(x_1, x_2)u \, dx_1 dx_2 \approx d(P) U_P \times \text{(area of } S)$$

$$= \frac{h^2}{4} d(P) U_P (s_1 + s_3)(s_2 + s_4). \tag{36}$$

Collecting the terms in (34), (35) and (36) leads to the five-point difference equation

$$\sum_{i=1}^{4} \alpha_i U_{Q_i} - \alpha_0 U_P = 0$$

with

$$\alpha_1 = \tfrac{1}{2}(s_2 + s_4)a_1(N_1)/s_1, \quad \alpha_2 = \tfrac{1}{2}(s_1 + s_3)a_2(N_2)/s_2$$
$$\alpha_3 = \tfrac{1}{2}(s_2 + s_4)a_1(N_3)/s_3, \quad \alpha_4 = \tfrac{1}{2}(s_1 + s_3)a_2(N_4)/s_4$$

and

$$\alpha_0 = \sum_{i=1}^{4} \alpha_i + \frac{1}{4}h^2 (s_1 + s_3)(s_2 + s_4)\mathrm{d}(P).$$

Exercise

5. Show that the finite difference approximation to (32) derived by the box integration scheme is identical to formula (27) given in section 3.6.

When the difference equations have to be solved in conjunction with Neumann type boundary conditions, it is natural to align the mesh so that the appropriate part of the boundary coincides with $C_1 C_2$, $C_2 C_3$, $C_3 C_4$ or $C_4 C_1$ (see figure 4). In this way the boundary condition can be incorporated directly into terms of the form (33).

The box integration scheme can be extended to more general non-rectilinear grids although, when Neumann or Robbins boundary conditions are prescribed, it may be necessary to restrict $a_2 = a_1$ in (32). When the box integration scheme is applicable it leads to a system of linear equations with a symmetric, diagonally dominant coefficient matrix whose diagonal entries are positive and off-diagonal entries non-positive. We shall see in subsequent sections that these are very desirable properties.

3.8 General properties of difference formulae

The preceding sections have described means by which an elliptic equation of the form

$$Lu = f \tag{37}$$

may be approximated by a difference equation

$$L_h U \equiv \alpha_0 U_P - \sum_{i=1}^{s} \alpha_i U_{Q_i} = f_P \tag{38}$$

where Q_i ($i = 1, 2, \ldots, s$) are grid points adjacent to P. It will generally be necessary to multiply some of our previous difference equations by suitable constants in order that the right hand sides of (37) and (38) agree. The construction of difference equations generally allows considerable scope in the determination of the coefficients α_i ($i = 0, 1, \ldots, s$). It is often convenient if these coefficients satisfy one or more of the following conditions:

(i)   *non-negativeness* of the coefficients

$$\left. \begin{array}{l} \alpha_0 > 0 \\ \\ \alpha_i \geqslant 0, i = 1, 2, \ldots, s \end{array} \right\} \text{ for each P.}$$

122

(ii)   *diagonal dominance* of the coefficient matrix

$$\sum_{i=1}^{s} |\alpha_i| \leqslant \alpha_0 \quad \text{(for all P)}$$

with strict inequality for at least one P.

(iii)   *symmetry* of the coefficient matrix.

Particularly with non-self-adjoint problems, all or some of these conditions may be violated as illustrated by the following exercise.

Exercises

6. The difference equations

$$4U_{l,m} - (1 - \tfrac{1}{2}kh)U_{l+1,m} - (1 + \tfrac{1}{2}kh)U_{l-1,m} - U_{l,m+1} - U_{l,m-1} = 0 \tag{39}$$

$$(4 + kh)U_{l,m} - U_{l+1,m} - (1 + kh)U_{l-1,m} - U_{l,m-1} - U_{l,m-1} = 0 \tag{40}$$

$$(2 + 2\cosh \tfrac{1}{2}kh)U_{l,m} - e^{-\tfrac{1}{2}kh}U_{l+1,m} - e^{\tfrac{1}{2}kh}U_{l-1,m} - U_{l,m+1} - U_{l,m-1} = 0 \tag{41}$$

are all consistent replacements of the elliptic equation

$$-\frac{\partial^2 u}{\partial x_1^2} - \frac{\partial^2 u}{\partial x_2^2} + k\frac{\partial u}{\partial x_1} = 0 \quad (k > 0), \tag{42}$$

on a square mesh of size h. Show that

(a)   Equation (39) has non-negative coefficients and is diagonally dominant only when $h < 2/k$.
(b)   Equations (40) and (41) have non-negative coefficients and are diagonally dominant for all values of h and k $(\geqslant 0)$.
(c)   Equation (41) leads to a symmetric coefficient matrix if it is first multiplied by $\exp(-lh)$ whereas (39) and (40) always lead to unsymmetric matrices.

7. Use equations (39)–(41) in turn to solve (42) on the unit square $0 \leqslant x_1, x_2 \leqslant 1$ with a grid of size $h = \tfrac{1}{4}$ for $k = 40$. The boundary conditions are $u(0, x_2) = \sin \pi x_2$ $(0 < x_2 < 1)$ and $u = 0$ on the remaining sides of the square.
[The exact solution of (42) subject to these boundary conditions is

$$u(x_1, x_2) = e^{\tfrac{1}{2}kx_1} \sinh \sigma(1 - x_1) \sin \pi x_2 / \sinh \sigma$$

where $\sigma = \tfrac{1}{2}(k^2 + 4\pi^2)^{\tfrac{1}{2}}$.
The computed values are

| Grid | Exact | Numerical solutions | | |
point	solution	Eqn 39	Eqn 40	Eqn 41
(¼, ¼)	0.6651	0.921	0.668	0.704
(½, ¼)	0.6255	0.466	0.627	0.702
(¾, ¼)	0.5883	1.080	0.544	0.699
(¼, ½)	0.9405	1.302	0.944	0.996
(½, ½)	0.8846	0.658	0.886	0.992
(¾, ½)	0.8320	1.528	0.769	0.988

The values at the grid points along $x_2 = ¾$ are the same as those along $x_2 = ¼$. The values computed by equation (39) are quite unacceptable because they violate the maximum principle.]

Symmetry of the coefficient matrix (condition (iii)) allows more efficient algorithms to be employed for the solution of the assembled equations. Satisfaction of conditions (i) and (ii) are, however, more fundamental since they ensure that the corresponding difference equation satisfies a *maximum principle* analogous to that stated for elliptic differential equations in section 3.1. More precisely, if

$$L_h V \leqslant 0 \tag{43}$$

at all grid points in R, then the value of V at grid points inside R nowhere exceeds its maximum on the boundary ∂R of R.

Example 4. Show that the maximum principle holds for the difference operator L_h defined by (38) provided conditions (i) and (ii) are satisfied.

Suppose that all grid values of V are positive. This is easily ensured by adding a suitably large positive constant to all values of V throughout the field. Now

$$\alpha_0 V_P \leqslant \sum_{i=1}^{s} \alpha_i V_{Q_i} \qquad \text{(by (43))}$$

$$\leqslant \max_i V_{Q_i} (\sum_{i=1}^{s} \alpha_i) \qquad \text{(from (i))}$$

$$\leqslant \max_i V_{Q_i} \alpha_0 \qquad \text{(from (ii))}$$

and so

$$V_P \leqslant \max_i V_{Q_i} .$$

Thus the value of V at the grid point P does not exceed the largest value at neighbouring grid points $Q_1, Q_2, \ldots, Q_s$. The same argument applies at all neighbours of the original point, until a grid point with neighbours on ∂R is reached. Hence the required result is proved.

To illustrate some of the implications of the maximum principle, we consider the special case

$$Lu \equiv -\left(\frac{\partial^2 u}{\partial x_1^2} + \frac{\partial^2 u}{\partial x_2^2}\right) = f \tag{44}$$

in the unit square $R = [\![0 \leqslant x_1, x_2 \leqslant 1]\!]$ with Dirichlet boundary conditions $u = g$ on ∂R. Using the familiar five-point difference replacement of (44) gives

$$L_h U \equiv \frac{1}{h^2} [4U_{l,m} - U_{l+1,m} - U_{l-1,m} - U_{l,m+1} - U_{l,m-1}] = f_{l,m} \tag{45}$$

for $1 \leqslant l,m \leqslant M - 1$. By means of Taylor expansions (see (5)) we deduce that

$$|L_h u - Lu| \leqslant \tau = \tfrac{1}{6} h^2 M_4 \tag{46}$$

where

$$M_4 = \max_{0 < x_1, x_2 < 1} \left\{ \left|\frac{\partial^4 u}{\partial x_1^4}\right|, \left|\frac{\partial^4 u}{\partial x_2^4}\right| \right\}$$

provided $u \in C^{4,4}$. Moreover, since

$$(Lu)_{l,m} = f_{l,m} = (L_h U)_{l,m} ,$$

equation (46) may be written as

$$|L_h e| \leqslant \tau \tag{47}$$

where $e_{l,m} = u_{l,m} - U_{l,m}$ is the error at the grid point (lh, mh). We shall now demonstrate how the maximum principle can be employed to derive an upper bound on the error e. Introducing the function $\phi = \tfrac{1}{2}x_1^2$, a simple calculation shows that

$$L_h \phi = -1. \tag{48}$$

In view of (47) and (48)

$$L_h(e + \tau\phi) = L_h e + \tau L_h \phi$$
$$= L_h e - \tau \leqslant 0$$

and, by appealing to the maximum principle, we find that

$$e + \tau\phi \leqslant \max_{\partial R} (e + \tau\phi) = \tfrac{1}{2}\tau$$

since e vanishes on ∂R. Further, the positivity of ϕ in R allows us to conclude that

$$e \leqslant \tfrac{1}{2}\tau.$$

Repeating these steps with the function $(-e + \tau\phi)$ shows that $-e \leqslant \tfrac{1}{2}\tau$ and consequently

$$|e_{l,m}| = |u_{l,m} - U_{l,m}| \leqslant \tfrac{1}{2}\tau = \tfrac{1}{12}h^2 M_4 \ . \tag{49}$$

Therefore the local truncation error τ and the error e are both $0(h^2)$ and the numerical solution therefore converges to the exact solution as $h \to 0$.

It is possible to maintain the $0(h^2)$ bound on the error e even on irregular regions R, subject to mixed boundary conditions on ∂R, provided

(a) The difference equation satisfies conditions (i) and (ii) at each grid point.
(b) The boundary conditions are approximated with an error not exceeding $0(h^2)$.
(c) The local truncation error of difference replacements at grid points adjacent to ∂R is at least $0(h)$.
(d) The local truncation error at other internal grid points of R is at least $0(h^2)$.
(e) The fourth derivatives of u exist and are bounded in R.

Exercise

8. Show that the treatment of the Dirichlet problem for Laplace's equation described in section 3.5 satisfies the requirements (a)–(d) listed above.

The fact that the presence of less accurate difference approximations near the boundary does not appear to diminish the accuracy is of great practical importance. This is demonstrated by comparing the results of Exercise 2 and Example 2.

The treatment described in section 3.5 of Neumann type boundary conditions on curved boundaries has an accuracy of only $0(h)$. In view of (b) above, the error at internal grid points should not therefore be expected to be less than $0(h)$.

3.9 Mixed partial derivatives

So far the elliptic equations considered in this chapter have not contained any mixed partial derivatives. Accordingly, we now examine the solution, using finite difference methods, of the elliptic partial differential equation

$$Lu \equiv a(x_1, x_2)\frac{\partial^2 u}{\partial x_1^2} + 2b(x_1, x_2)\frac{\partial^2 u}{\partial x_1 \partial x_2} + c(x_1, x_2)\frac{\partial^2 u}{\partial x_2^2} = 0 \tag{50}$$

in the closed region R, where the ellipticity condition gives

$$ac > b^2, \quad (x_1, x_2) \in R.$$

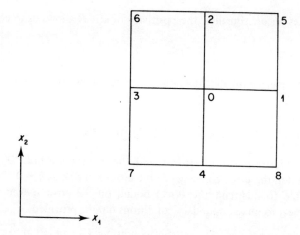

Figure 5

A square grid, with grid size h, is now placed over the region R in the most convenient manner, and a typical internal grid point O is represented in figure 5. The partial derivatives $\partial^2 u/\partial x_1^2$ and $\partial^2 u/\partial x_2^2$ are replaced as usual by

$$\frac{1}{h^2}(u_1 - 2u_0 + u_3)$$

and

$$\frac{1}{h^2}(u_2 - 2u_0 + u_4)$$

respectively. The mixed partial derivative $\partial^2 u/\partial x_1 \partial x_2$ requires careful consideration and initially it is replaced by

$$\frac{1}{h^2}[\alpha_1(u_5 + u_0 - u_2 - u_1) + \alpha_2(u_2 + u_3 - u_6 - u_0) +$$

$$\alpha_3(u_0 + u_7 - u_3 - u_4) + \alpha_4(u_1 + u_4 - u_0 - u_8)], \tag{51}$$

where

$$\sum_{i=1}^{4} \alpha_i = 1. \tag{52}$$

Taylor expansions about 0 give

$$\frac{1}{h^2}(u_5 + u_0 - u_2 - u_1) = \frac{\partial^2 u}{\partial x_1 \partial x_2} + \tfrac{1}{2}h\left(\frac{\partial^3 u}{\partial x_1^2 \partial x_2} + \frac{\partial^3 u}{\partial x_1 \partial x_2^2}\right) + \dots,$$

$$\frac{1}{h^2}(u_2 + u_3 - u_6 - u_0) = \frac{\partial^2 u}{\partial x_1 \partial x_2} + \tfrac{1}{2}h\left(-\frac{\partial^3 u}{\partial x_1^2 \partial x_2} + \frac{\partial^3 u}{\partial x_1 \partial x_2^2}\right) + \ldots ,$$

$$\frac{1}{h^2}(u_0 + u_7 - u_3 - u_4) = \frac{\partial^2 u}{\partial x_1 \partial x_2} + \tfrac{1}{2}h\left(-\frac{\partial^3 u}{\partial x_1^2 \partial x_2} - \frac{\partial^3 u}{\partial x_1 \partial x_2^2}\right) + \ldots ,$$

$$\frac{1}{h^2}(u_1 + u_4 - u_0 - u_8) = \frac{\partial^2 u}{\partial x_1 \partial x_2} + \tfrac{1}{2}h\left(\frac{\partial^3 u}{\partial x_1^2 \partial x_2} - \frac{\partial^3 u}{\partial x_1 \partial x_2^2}\right) + \ldots ,$$

and so equation (51) becomes

$$\frac{\partial^2 u}{\partial x_1 \partial x_2} + \tfrac{1}{2}h(\alpha_1 - \alpha_2 - \alpha_3 + \alpha_4)\frac{\partial^3 u}{\partial x_1^2 \partial x_2} + \tag{53}$$

$$\tfrac{1}{2}h(\alpha_1 + \alpha_2 - \alpha_3 - \alpha_4)\frac{\partial^3 u}{\partial x_1 \partial x_2^2} + \ldots \; .$$

The terms in $\partial^3 u/\partial x_1^2 \partial x_2$ and $\partial^3 u/\partial x_1 \partial x_2^2$ vanish if

and
$$\left.\begin{array}{c}(\alpha_1 - \alpha_3) - (\alpha_2 - \alpha_4) = 0 \\[2mm] (\alpha_1 - \alpha_3) + (\alpha_2 - \alpha_4) = 0. \end{array}\right\} \tag{54}$$

If the coefficients α_i ($i = 1, 2, 3, 4$) are chosen to satisfy the system of equations (52) and (54), then a difference approximation to equation (50) can be constructed with a local truncation error $0(h^2)$.

Exercise

9. Show that if $\alpha_1 = \alpha_2 = \alpha_3 = \alpha_4 = \tfrac{1}{4}$, the principal part of the local truncation error is

$$\tfrac{1}{6}h^2 \frac{\partial^4 u}{\partial x_1^2 \partial x_2^2} \; .$$

Returning to equation (50) and replacing the derivatives as suggested, the difference equation

$$2[a + c + b(-\alpha_1 + \alpha_2 - \alpha_3 + \alpha_4)] U_0 =$$
$$[a + 2b(\alpha_4 - \alpha_1)] U_1 + [c + 2b(\alpha_2 - \alpha_1)] U_2 +$$
$$[a + 2b(\alpha_2 - \alpha_3)] U_3 + [c + 2b(\alpha_4 - \alpha_3)] U_4 +$$
$$2b\alpha_1 U_5 - 2b\alpha_2 U_6 + 2b\alpha_3 U_7 - 2b\alpha_4 U_8 \tag{55}$$

128

is obtained. In the previous section, it was suggested that a desirable feature of a difference formula replacing an elliptic differential equation is that its coefficients be non-negative. We now impose this condition on the coefficients in equation (55) and two cases are considered. In each case, since the ellipticity condition requires a and c to be of the same sign, no loss of generality occurs if we choose $a(x_1, x_2) > 0$, $c(x_1, x_2) > 0$ for all $(x_1, x_2) \in R$.

Case I $\quad b(x_1, x_2) > 0$

It follows immediately from equation (55) that

$$\alpha_1 \geqslant 0, \ \alpha_2 \leqslant 0, \ \alpha_3 \geqslant 0, \ \alpha_4 \leqslant 0,$$

where all four equalities cannot hold. A convenient set of values which satisfies these inequalities and also satisfies equations (52) and (54) is

$$\alpha_1 = \alpha_3 = \tfrac{1}{2}; \quad \alpha_2 = \alpha_4 = 0.$$

With these values, equation (55) becomes

$$2(a + c - b)U_0 = (a - b)(U_1 + U_3) + \\ (c - b)(U_2 + U_4) + b(U_5 + U_7), \tag{56}$$

which has positive coefficients if

$$0 < b < \min [a,c].$$

Case II $\quad b(x_1, x_2) < 0$

This time, from equation (55) it follows that

$$\alpha_1 \leqslant 0, \ \alpha_2 \geqslant 0, \ \alpha_3 \leqslant 0, \ \alpha_4 \geqslant 0,$$

where all four equalities cannot hold, and we choose

$$\alpha_1 = \alpha_3 = 0; \quad \alpha_2 = \alpha_4 = \tfrac{1}{2},$$

leading to equation (55) in the form

$$2(a + c + b)U_0 = (a + b)(U_1 + U_3) + (c + b)(U_2 + U_4) - b(U_6 + U_8), \tag{57}$$

which has positive coefficients if

$$0 < -b < \min [a,c].$$

Thus, irrespective of the sign of b, values of α_i ($i = 1, 2, 3, 4$) can be found which make the coefficients of equation (55) positive, provided

$$|b| < \min [a,c],$$

a condition which was found by Greenspan and Jain (1964).

Unfortunately, there can be a big difference between $|b| < (ac)^{1/2}$, the condition for ellipticity, and $|b| < \min [a,c]$, the condition for positive coefficients. For example, if $a = 2$, $c = 50$, the former condition gives $-10 < b < +10$, whereas the latter gives $-2 < b < +2$. The reader who wishes to pursue this point further is referred to Greenspan and Jain (1964), where it is shown that the condition for positive coefficients can be eliminated if the grid is generalized to a suitable rectangular form.

It should be noted that the matrix system based on either equation (56) or equation (57) at internal grid points of a closed region R suitably modified by Dirichlet boundary conditions involves a matrix which has diagonal dominance.

Exercise

10. Obtain the matrix system representing equation (50) in a square region with Dirichlet boundary conditions, and show that the matrix A has diagonal dominance.

3.10 The biharmonic equation

Preceding sections of this chapter have dealt only with the approximation of second-order elliptic problems and so we turn our attention to equations of the fourth order. As a model for discussion we consider the biharmonic equation

$$\frac{\partial^4 u}{\partial x_1^4} + 2\frac{\partial^4 u}{\partial x_1^2 \partial x_2^2} + \frac{\partial^4 u}{\partial x_2^4} = 0 \tag{58}$$

in a bounded region R. If $\partial/\partial n$ denotes differentiation in an outward normal direction to the boundary ∂R, the appropriate boundary conditions usually take one of two forms

I $\qquad u = f(x_1, x_2)$ $\qquad\qquad\qquad\qquad\qquad\qquad\qquad$ (58a)

$\qquad\qquad \dfrac{\partial u}{\partial n} = g(x_1, x_2)$ $\left.\right\}$ for $(x_1, x_2) \in \partial R$ $\qquad$ (58b)

or

II $\qquad u = f(x_1, x_2)$ $\qquad\qquad\qquad\qquad\qquad\qquad\qquad$ (59a)

$\qquad\qquad \dfrac{\partial^2 u}{\partial n^2} = g(x_1, x_2)$ $\left.\right\}$ for $(x_1, x_2) \in \partial R$. $\qquad$ (59b)

To simplify the ensuing discussion we shall assume that R is the unit square $0 \leqslant x_1, x_2 \leqslant 1$. An obvious replacement of (58) on a square grid of size h is given by

$$\delta_{x_1}^4 U_{l,m} + 2\delta_{x_1}^2 \delta_{x_2}^2 U_{l,m} + \delta_{x_2}^4 U_{l,m} = 0. \tag{60}$$

This is a 13-point difference equation involving grid values of U at the nodes depicted by $\times$ in figure 6. Written out explicitly, (60) becomes

130

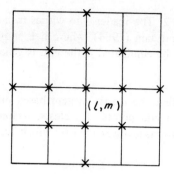

Figure 6

$$20U_{l,m} - 8(U_{l+1,m} + U_{l-1,m} + U_{l,m+1} + U_{l,m-1})$$
$$+ 2(U_{l+1,m+1} + U_{l+1,m-1} + U_{l-1,m+1} + U_{l-1,m-1})$$
$$+ U_{l+2,m} + U_{l-2,m} + U_{l,m+2} + U_{l,m-2} = 0. \tag{61}$$

The presence of the terms $U_{l\pm2,m}$ and $U_{l,m\pm2}$ in this equation requires special provision to be made when it is applied at nodes a distance h from the boundary. This involves extending the grid outside R in the manner shown in figure 7 for $h = \frac{1}{4}$. Equation (61) can now be applied to all internal grid points of R and the values outside R are eliminated by suitable approximations of the boundary conditions. The process will be discussed for the grid shown in figure 7.

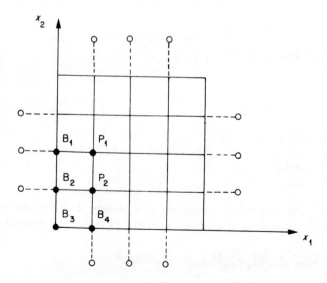

Figure 7

Boundary conditions I

When (61) is applied at the node $(h, 2h)$ (P_1 in figure 7) it becomes

$$20U_{1,2} - 8(U_{2,2} + U_{0,2} + U_{1,3} + U_{1,1}) + 2(U_{2,3} + U_{2,1} + U_{0,3} + U_{0,1})$$
$$+ U_{3,2} + U_{-1,2} + U_{1,4} + U_{1,0} = 0. \tag{62}$$

From the boundary condition (58a)

$$U_{0,1} = f_{0,1}, \quad U_{0,2} = f_{0,2}, \quad U_{0,3} = f_{0,3}, \quad \text{and} \quad U_{1,0} = f_{1,0}. \tag{63}$$

Further, since $\partial u/\partial n \equiv -\partial u/\partial x_1$ on $x_1 = 0$, (58b) is replaced by

$$- (U_{1,2} - U_{-1,2})/2h = g_{0,2}. \tag{64}$$

Combining (62), (63) and (64) there follows

$$21U_{1,2} - 8(U_{2,2} + U_{1,3} + U_{1,1}) + 2(U_{2,3} + U_{2,1}) + U_{3,2} + U_{1,4} =$$
$$8f_{0,2} - 2(f_{0,3} + f_{0,1}) - f_{1,0} - 2hg_{0,2}, \tag{65}$$

in which the left hand side involves only grid values of U lying inside R and the right hand side consists of known function values.

A similar procedure is adopted for all nodes adjacent to the boundary. For corner points such as P_2 in figure 7 approximations to the boundary condition must be invoked at both B_2 and B_4.

Boundary conditions II

The only modification to the above strategy is to replace (64) by an approximation to the boundary condition (59b) at B_1 $(0, 2h)$. Using the fact that

$$\left(\frac{\partial^2 u}{\partial x_1^2}\right)_{B_1} \approx (U_{1,2} - 2U_{0,2} + U_{-1,2})/h^2,$$

we have

$$U_{1,2} - 2U_{0,2} + U_{-1,2} = h^2 g_{0,2}. \tag{66}$$

Eliminating $U_{-1,2}$ between (62) and (66) and substituting the boundary values (63) we have, in place of (65)

$$19U_{1,2} - 8(U_{2,2} + U_{1,3} + U_{1,1}) + 2(U_{2,3} + U_{2,1}) + U_{3,2} + U_{1,4} =$$
$$6f_{0,2} - 2(f_{0,3} + f_{0,1}) - f_{1,0} - h^2 g_{0,2}.$$

The extension of these ideas to non-rectangular regions is described by Zlamal (1967) who also proves that, when u is sufficiently smooth, the error $u - U$ at grid points is $0(h^2)$.

In view of our earlier discussion in section 3.8 on desirable properties of difference equations, it should be noted that (61) is neither diagonally dominant nor are its coefficients non-negative. However, when the difference equations are assembled in matrix form, the coefficient matrix is positive definite. Other difference replacements of (58) which may be more appropriate for iterative methods of solution are developed by Tee (1963).

To conclude this section we mention an alternative strategy for solving the biharmonic equation. This is based on the observation that (58) may be written in the form

$$\left(\frac{\partial^2}{\partial x_1^2} + \frac{\partial^2}{\partial x_2^2}\right)\left(\frac{\partial^2 u}{\partial x_1^2} + \frac{\partial^2 u}{\partial x_2^2}\right) = 0 \tag{67}$$

and by defining

$$\frac{\partial^2 u}{\partial x_1^2} + \frac{\partial^2 u}{\partial x_2^2} = v, \tag{68}$$

(67) becomes

$$\frac{\partial^2 v}{\partial x_1^2} + \frac{\partial^2 v}{\partial x_2^2} = 0. \tag{69}$$

We are led therefore to a coupled pair of second-order elliptic equations which may be replaced by standard five-point difference approximations. Before these equations can be solved, the boundary conditions on u must be approximated in such a way as to also provide boundary values for v. This involves an iterative process a full description of which is beyond the scope of this book. The interested reader is referred to the works of Pearson (1965), Smith (1968) and McLaurin (1974).

Example 5. Show that when (68) and (69) are to be solved on the unit square $R = [\![0 \leqslant x_1, x_2 \leqslant 1]\!]$ subject to the boundary conditions (59a) and (59b), explicit values for both u and v may be determined on ∂R.

Consider the portion of the boundary $x_1 = 0, 0 < x_2 < 1$. By (59a)

$$u(0, x_2) = f(0, x_2)$$

and therefore

$$\frac{\partial^2 u}{\partial x_2^2}(0, x_2) = \frac{\partial^2 f}{\partial x_2^2}(0, x_2).$$

The boundary condition (59b) gives

$$\frac{\partial^2 u}{\partial x_1^2}(0, x_2) = g(0, x_2)$$

and, since (68) implies that

$$v(0, x_2) = \left(\frac{\partial^2 u}{\partial x_1^2} + \frac{\partial^2 u}{\partial x_2^2} \right)_{(0, x_2)}$$

we conclude that

$$v(0, x_2) = g(0, x_2) + \frac{\partial^2 f}{\partial x_2^2} (0, x_2).$$

A similar argument shows that v may be determined along each side of ∂R and consequently (69) may be solved subject to Dirichlet type boundary conditions. When this solution is substituted into the right hand side of (68), this equation can be solved subject to (59a). Unfortunately, this approach is not possible when the boundaries are not parallel to the co-ordinates axes or when boundary conditions of type I are imposed.

3.11 The solution of elliptic difference equations

In preceding sections it has been shown how difference replacements of elliptic differential equations lead to systems of linear equations of the form

$$A\mathbf{x} = \mathbf{b} \tag{70}$$

in which the $N \times N$ coefficient matrix A has relatively few non-zero entries in each row and these form a well-defined pattern. Such matrices are said to be *sparse* and efficient routines for solving (70) have to be capable of exploiting this property. Broadly speaking, possible methods fall into one of two categories, namely *direct* and *iterative*. Methods of the former type produce a solution to (70) in a finite number of operations which can be predicted in advance. In contrast to this, iterative methods require an initial approximation and thereafter generate a sequence of vectors which, under favourable conditions, converge to the exact solution. Since the number of arithmetic operations involved depends on the initial approximation and the criterion for terminating the iteration, as well as other factors, it cannot be predicted in advance. Iterative methods do however take full account of sparsity, unlike direct methods which are only partially successful in this respect. Consequently, iterative methods are usually preferred when the order N of the system is large.

3.12 Direct factorization methods

We begin by describing a method for factorizing the $N \times N$ coefficient matrix A into a product LU where U and L are *upper* and *unit lower* triangular matrices respectively. Writing

$$L = \begin{bmatrix} 1 & & & & & & \\ l_{21} & 1 & & & & \bigcirc & \\ l_{31} & l_{32} & 1 & & & & \\ \cdot & \cdot & \cdot & \cdot & & & \\ \cdot & & & & \cdot & & \\ \cdot & & & & & & \\ l_{N1} & \cdot & \cdot & \cdot & \cdot & \cdot & 1 \end{bmatrix} \quad \text{and} \quad U = \begin{bmatrix} u_{11} & u_{12} & u_{13} & \cdot & \cdot & \cdot & u_{1N} \\ & u_{22} & u_{23} & \cdot & \cdot & \cdot & u_{2N} \\ & & \cdot & & & & \cdot \\ & & & \cdot & & & \\ & \bigcirc & & & \cdot & & \\ & & & & & \cdot & \\ & & & & & & u_{NN} \end{bmatrix}$$

the N^2 unknown elements of these two matrices are computed so that

$$LU = A. \tag{72}$$

This can be achieved by computing each element in the product LU and equating the result with the corresponding element of A. In this way it can be shown that the non-zero entries in the ith row of U are given by

$$u_{ij} = a_{ij} - \sum_{k=1}^{i-1} l_{ik} u_{kj} \quad (j = i, i+1, \ldots, N), \tag{73}$$

and those in the ith column of L (below the diagonal) by

$$l_{ji} = (a_{ji} - \sum_{k=1}^{i-1} l_{jk} u_{ki})/u_{ii} \quad (j = i+1, i+2, \ldots, N), \tag{74}$$

for $i = 1, 2, \ldots, N$. It is understood that the summations in (73), (74) and the remainder of this section should be ignored when the lower limit of the index k exceeds the upper limit. It is clear from (74) that the factorization cannot be completed if any diagonal entry in the matrix U is zero. This possibility may be circumvented by partial pivoting which allows a permutation of rows of A. Fortunately, matrices arising from finite difference approximations to elliptic equations are usually either diagonally dominant or symmetric positive definite in which case the LU factorization exists without recourse to pivoting.

Having completed the factorization of A, the linear equations $Ax = b$ may be decomposed into the triangular systems

$$Ly = b \tag{75}$$

and

$$Ux = y. \tag{76}$$

These are easily solved by forward and backward substitution to give

$$y_i = b_i - \sum_{k=1}^{i-1} l_{ik} y_k \quad (i = 1, 2, \ldots, N), \tag{77}$$

and

$$x_i = (y_i - \sum_{k=i+1}^{N} u_{ik} x_k)/u_{ii} \quad (i = N, N-1, \ldots, 1). \tag{78}$$

where again it is understood that the summations are ignored if the lower limit of the index exceeds the upper limit.

Exercise

11. Show that the LU factors of the matrix

$$A = \begin{bmatrix} 2 & -1 & -2 \\ -4 & 3 & 1 \\ -2 & 0 & 9 \end{bmatrix}$$

are

$$L = \begin{bmatrix} 1 & 0 & 0 \\ -2 & 1 & 0 \\ -1 & -1 & 1 \end{bmatrix} \quad \text{and} \quad U = \begin{bmatrix} 2 & -1 & -2 \\ 0 & 1 & -3 \\ 0 & 0 & 4 \end{bmatrix}.$$

Use these to solve the system $A\mathbf{x} = \mathbf{b}$ where $\mathbf{b} = \{1, -5, 10\}^T$.

LU factorization of band matrices

The matrix A is said to be a *band* matrix of *band width* $2p + 1$ if $a_{ij} = 0$ for $|i - j| > p$. Obviously A will have at most $2p + 1$ non-zero entries in each row and, when $p = 1$ for instance, A is a tridiagonal matrix.

Applying the above LU factorization process to a band matrix A we find that

$$l_{ij} = 0 \quad \text{for} \quad j > i \text{ and } j < i - p,$$

and

$$u_{ij} = 0 \quad \text{for} \quad i > j \text{ and } i < j - p.$$

That is to say, L and U are also band matrices. The remaining non-zero elements of L and U are given by

$$u_{ij} = a_{ij} - \Sigma l_{ik} u_{kj} \quad (j = i, i + 1, \ldots, i + p), \tag{79}$$

and

$$l_{ji} = (a_{ji} - \Sigma l_{jk} u_{ki})/u_{ii} \quad (j = i + 1, i + 2, \ldots, i + p), \tag{80}$$

for $i = 1, 2, \ldots, N$ where the summations extend from $k = \max(1, j - p)$ to $k = i - 1$. The solution $\mathbf{x}$ is again calculated using (77) and (78) with the summations ranging from $k = \max(1, i - p)$ to $k = i - 1$ and from $k = i + 1$ to $k = \min(N, i + p)$ respectively. This modified procedure involves significantly less computation when $p \ll N$.

136

Coefficient matrices obtained from finite difference approximations to second order elliptic equations on a grid of size $h = 1/M$ have dimension $0(M^2)$ and typically have only five non-zero entries in each row. On the other hand the triangular factors L and U have $0(M)$ non-zero entries in each row. The presence of the additional non-zero elements is referred to as *fill-in* and much recent progress has been made towards devising orderings of the equations so that it is reduced as far as possible. See for instance the nested dissection method of George described in George, Poole and Voigt (1978).

Exercise

12. Let A denote the matrix obtained from the usual five-point difference replacement of Laplace's equation on the unit square subject to Dirichlet boundary conditions. If the grid size is $h = 1/M$, show that A has dimension $N = (M - 1)^2$ and, for the natural ordering of gridpoints, has band width $2M - 1$.

Factorization of symmetric positive definite matrices

A significant reduction in the amount of computation can also be achieved for both the above factorization methods when A is symmetric and positive definite. Defining the diagonal matrix

$$D = \begin{bmatrix} u_{11} & & & \\ & u_{22} & & \\ & & \ddots & \\ & & & u_{NN} \end{bmatrix}$$

then A may be factorized into the product

$$A = LDL^T. \tag{81}$$

Proceeding as for the general LU factorization we find that

$$u_{ii} = a_{ii} - \sum_{k=1}^{i-1} l_{ik}^2, \tag{82}$$

and

$$l_{ji} = (a_{ji} - \sum_{k=1}^{i-1} l_{jk} l_{ik})/u_{ii} \quad (j = i+1, \ldots, N), \tag{83}$$

for $i = 1, 2, \ldots, N$. The process described is closely related to the Cholesky factorization of a symmetric positive definite matrix. The work involved is approximately one half of that for a general matrix since it is not necessary to compute (or store) elements above the diagonal in U. The positive-definiteness of A ensures that $u_{ii} > 0, i = 1, 2, \ldots, N$.

Corresponding to (77) and (78), the forward- and back-substitution stages are given by

$$y_i = b_i - \sum_{k=1}^{i-1} l_{ik} y_k \quad (i = 1, 2, \ldots, N), \tag{84}$$

and

$$x_i = y_i/u_{ii} - \sum_{k=i+1}^{N} l_{ki} x_k \quad (i = N, N-1, \ldots, 1). \tag{85}$$

The formulae (82)–(85) can again be adapted to the solution involving band matrices.

A more comprehensive treatment of factorization methods, including the effects of rounding errors, may be found in many specialized text books such as Broyden (1975), Stewart (1973) or Wilkinson (1965).

3.13 Successive overrelaxation (S.O.R.)

Full account can be taken of the sparsity pattern of the coefficient matrix by adopting iterative schemes. We assume that the approximation of an elliptic equation by finite differences leads to a set of simultaneous linear equations of the form

$$A'x = b'$$

in which A' is a non-singular square matrix with no zeros on its main diagonal. In order to simplify the algebra which is to follow, we rewrite this equation as

$$Ax = b \quad (|A| \neq 0) \tag{86}$$

where $A = DA'$, $b = Db'$ and D is a diagonal matrix with elements $d_{ii} = 1/a_{ii}$, $i = 1, 2, \ldots, N$. The elements on the principal diagonal of A are unity and we can thus split A into the form

$$A = I - L - U$$

where I is the unit matrix and L, U are strictly lower and strictly upper triangular matrices respectively. The matrices L and U have zero diagonal elements and should not be confused with the LU factors of the previous section.

For example, the matrix defined by (7) and (8) with $M = 3$ gives

$$A = \begin{bmatrix} 1 & -\frac{1}{4} & -\frac{1}{4} & 0 \\ -\frac{1}{4} & 1 & 0 & -\frac{1}{4} \\ -\frac{1}{4} & 0 & 1 & -\frac{1}{4} \\ 0 & -\frac{1}{4} & -\frac{1}{4} & 1 \end{bmatrix},$$

$$L = \begin{bmatrix} 0 & 0 & 0 & 0 \\ -\frac{1}{4} & 0 & 0 & 0 \\ -\frac{1}{4} & 0 & 0 & 0 \\ 0 & -\frac{1}{4} & -\frac{1}{4} & 0 \end{bmatrix} \quad \text{and} \quad U = \begin{bmatrix} 0 & -\frac{1}{4} & -\frac{1}{4} & 0 \\ 0 & 0 & 0 & -\frac{1}{4} \\ 0 & 0 & 0 & -\frac{1}{4} \\ 0 & 0 & 0 & 0 \end{bmatrix}.$$

138

An iterative process is now set up to solve equation (86). This will succeed if an initial guess x_0 is successively improved by the iterative process until it is arbitrarily close to the solution x. The process is represented diagrammatically by

$$x_0 \to x_1 \to x_2 \ldots \to x_i \to x_{i+1} \to \ldots \to x.$$

If we write equation (86) in the form

$$(I - L - U)x = b, \tag{87}$$

and x_i and x_{i+1} are successive approximate solutions of equation (86), then equation (87) can be used to give

$$Ix_{i+1} = (L + U)x_i + b \quad (i = 0, 1, 2, \ldots), \tag{88}$$

which is the *point Jacobi* iterative method, or

$$(I - L)x_{i+1} = Ux_i + b, \tag{89}$$

which is the *Gauss–Seidel* iterative method. These two methods are particular cases of the general iterative process

$$x_{i+1} = Bx_i + c \quad (i = 0, 1, 2, \ldots), \tag{90}$$

where $B = L + U$ and $(I - L)^{-1}U$ in the point Jacobi and Gauss–Seidel processes respectively. If x is the desired solution, satisfying

$$x = Bx + c,$$

and the error in the ith iterate is

$$e_i = x_i - x \quad (i = 0, 1, 2, \ldots),$$

then

$$e_{i+1} + x = Be_i + Bx + c$$

and so

$$e_{i+1} = Be_i \quad (i = 0, 1, 2, \ldots).$$

It follows that

$$e_i = B^i e_0,$$

and so

$$e_i \to 0 \text{ as } i \to \infty, \text{ if } B^i \to 0,$$

where 0 is the null matrix. B is *convergent* (i.e. $B^i \to 0$ as $i \to \infty$) if and only if $\rho(B) < 1$ (see chapter 1, page 15). Thus the iterative process (90) is convergent if and only if $\rho(B) < 1$. [$\rho(B)$ is the maximum modulus eigenvalue or spectral radius of the matrix B.]

Example 6. Calculate the spectral radii $\rho(L + U)$ and $\rho[(I - L)^{-1}U]$ of the point Jacobi and Gauss–Seidel methods respectively as applied to the solution of the system of equations

$$
\begin{aligned}
-2x + y &= 0 \\
x - 2y + z &= 0 \\
y - 2z &= -4,
\end{aligned}
$$

and show that

$$\rho[(I - L)^{-1} U] = \rho^2 [L + U] < 1.$$

The equations can be written in the matrix form

$$
\begin{bmatrix} 1 & -\frac{1}{2} & 0 \\ -\frac{1}{2} & 1 & -\frac{1}{2} \\ 0 & -\frac{1}{2} & 1 \end{bmatrix}
\begin{bmatrix} x \\ y \\ z \end{bmatrix} =
\begin{bmatrix} 0 \\ 0 \\ 2 \end{bmatrix},
$$

from which

$$
L = \begin{bmatrix} 0 & 0 & 0 \\ \frac{1}{2} & 0 & 0 \\ 0 & \frac{1}{2} & 0 \end{bmatrix}, \quad
U = \begin{bmatrix} 0 & \frac{1}{2} & 0 \\ 0 & 0 & \frac{1}{2} \\ 0 & 0 & 0 \end{bmatrix}
$$

$$
I - L = \begin{bmatrix} 1 & 0 & 0 \\ -\frac{1}{2} & 1 & 0 \\ 0 & -\frac{1}{2} & 1 \end{bmatrix}, \quad
(I - L)^{-1} = \begin{bmatrix} 1 & 0 & 0 \\ \frac{1}{2} & 1 & 0 \\ \frac{1}{4} & \frac{1}{2} & 1 \end{bmatrix}
$$

and so

$$
(I - L)^{-1} U = \begin{bmatrix} 0 & \frac{1}{2} & 0 \\ 0 & \frac{1}{4} & \frac{1}{2} \\ 0 & \frac{1}{8} & \frac{1}{4} \end{bmatrix}, \quad
L + U = \begin{bmatrix} 0 & \frac{1}{2} & 0 \\ \frac{1}{2} & 0 & \frac{1}{2} \\ 0 & \frac{1}{2} & 0 \end{bmatrix}
$$

Eigenvalues of $(I - L)^{-1} U$ are given by

$$
\begin{vmatrix} -\lambda & \frac{1}{2} & 0 \\ 0 & \frac{1}{4} - \lambda & \frac{1}{2} \\ 0 & \frac{1}{8} & \frac{1}{4} - \lambda \end{vmatrix} = 0
$$

i.e. $\lambda = 0, 0, \frac{1}{2}$, and so

$$\rho[(I - L)^{-1} U] = \frac{1}{2}.$$

Eigenvalues of $(L + U)$ are given by

$$\begin{vmatrix} -\mu & \frac{1}{2} & 0 \\ \frac{1}{2} & -\mu & \frac{1}{2} \\ 0 & \frac{1}{2} & -\mu \end{vmatrix} = 0$$

i.e. $\mu = -1/\sqrt{2},\ 0,\ +1/\sqrt{2}$, and so

$$\rho(L + U) = \frac{1}{\sqrt{2}}$$

Thus both methods converge and

$$\rho[(I - L)^{-1}U] = \rho^2(L + U) < 1.$$

Exercise

13. Apply the Jacobi and Gauss–Seidel methods in the above example using the initial approximation $\mathbf{x}_0 = \{2, -3, 2\}^T$. Show that Jacobi iteration requires approximately twice as many iterations as the Gauss–Seidel method in order that each component be within 1% of the exact solution $\mathbf{x} = \{1, 2, 3\}^T$.

It is fair to say at the present time that one of the most successful iterative methods for solving systems of linear equations is the method of successive over-relaxation (S.O.R.). Again we start with the system of equations

$$A\mathbf{x} = \mathbf{b} \quad (|A| \neq 0),$$

which is written in the form

$$(I - L - U)\mathbf{x} = \mathbf{b}.$$

Introduce $\tilde{\mathbf{x}}_{i+1}$ given by

$$\tilde{\mathbf{x}}_{i+1} = L\mathbf{x}_{i+1} + U\mathbf{x}_i + \mathbf{b}, \tag{91}$$

where

$$\mathbf{x}_{i+1} = \omega\tilde{\mathbf{x}}_{i+1} + (1 - \omega)\mathbf{x}_i, \tag{92}$$

with $\omega(>0)$ an arbitrary parameter, independent of i. Three distinct cases arise:

(i) $\omega > 1$ overrelaxation,
(ii) $\omega = 1$ Gauss–Seidel, and
(iii) $0 < \omega < 1$ underrelaxation.

Elimination of $\tilde{\mathbf{x}}_{i+1}$ between equations (91) and (92) leads to

$$(I - \omega L)\mathbf{x}_{i+1} = [\omega U + (1 - \omega)I]\mathbf{x}_i + \omega\mathbf{b},$$

and so

$$\mathbf{x}_{i+1} = (I - \omega L)^{-1}[\omega U + (1 - \omega)I]\mathbf{x}_i + \omega(I - \omega L)^{-1}\mathbf{b}. \tag{93}$$

This is the iterative method of S.O.R. and fits into the general iterative process (90) with $B \equiv (I - \omega L)^{-1} [\omega U + (1 - \omega)I]$. From the theory of the general case, the method of S.O.R. will be convergent if and only if

$$\rho[(I - \omega L)^{-1} \{\omega U + (1 - \omega)I\}] < 1.$$

In order to simplify the notation, we introduce

$$L_\omega = (I - \omega L)^{-1} \{\omega U + (1 - \omega)I\},$$

and if λ is an eigenvalue of L_ω, then

$$|L_\omega - \lambda I| = 0. \tag{94}$$

Our aim is to calculate the maximum eigenvalue of L_ω from equation (94) and to minimize this with respect to ω.

Before commencing this calculation, it is necessary to give a few definitions.

Definition I   A matrix is **two-cyclic** *if by a suitable permutation of its rows and corresponding columns, it can be written in the form*

$$\begin{bmatrix} I_1 & F \\ G & I_2 \end{bmatrix},$$

where I_1, I_2 are square unit matrices, and F, G are arbitrary rectangular matrices.

Example 7

$$\begin{bmatrix} 1 & c & 0 \\ a & 1 & c \\ 0 & a & 1 \end{bmatrix} \rightarrow \begin{bmatrix} 1 & c & 0 \\ 0 & a & 1 \\ a & 1 & c \end{bmatrix} \rightarrow \begin{bmatrix} 1 & 0 & c \\ 0 & 1 & a \\ a & c & 1 \end{bmatrix}.$$

Thus the original matrix is two-cyclic, after first interchanging the second and third rows, and then interchanging the second and third columns.

Definition II   A matrix is **weakly two-cyclic** *if by a suitable permutation of its rows and corresponding columns, it can be written in the form*

$$\begin{bmatrix} 0_1 & F \\ G & 0_2 \end{bmatrix},$$

where 0_1, 0_2 are square null matrices.

Definition III   If a matrix $(I - L - U)$ is two-cyclic, then it is **consistently ordered** *if all the eigenvalues of the matrix*

$$\alpha L + \frac{1}{\alpha} U \quad (\alpha \neq 0)$$

are independent of α.

Example 8

$$\begin{bmatrix} 1 & -2 & 0 \\ -4 & 1 & -3 \\ 0 & -5 & 1 \end{bmatrix} = I - L - U$$

where

$$L = \begin{bmatrix} 0 & 0 & 0 \\ 4 & 0 & 0 \\ 0 & 5 & 0 \end{bmatrix}, \quad U = \begin{bmatrix} 0 & 2 & 0 \\ 0 & 0 & 3 \\ 0 & 0 & 0 \end{bmatrix}$$

$$\alpha L + (1/\alpha) U = \begin{bmatrix} 0 & 2/\alpha & 0 \\ 4\alpha & 0 & 3/\alpha \\ 0 & 5\alpha & 0 \end{bmatrix}$$

Eigenvalues λ of $\alpha L + (1/\alpha) U$ are given by

$$\begin{vmatrix} -\lambda & 2/\alpha & 0 \\ 4\alpha & -\lambda & 3/\alpha \\ 0 & 5\alpha & -\lambda \end{vmatrix} = 0,$$

i.e. $\lambda = 0, \pm\sqrt{23}$, independent of α, and so $I - L - U$ is consistently ordered.

Exercise

14. Show that the matrix

$$\begin{bmatrix} 1 & 0 & -\frac{1}{4} & -\frac{1}{4} \\ 0 & 1 & -\frac{1}{4} & -\frac{1}{4} \\ -\frac{1}{4} & -\frac{1}{4} & 1 & 0 \\ -\frac{1}{4} & -\frac{1}{4} & 0 & 1 \end{bmatrix}$$

is two-cyclic and consistently ordered.

Returning to the calculation of the maximum modulus eigenvalue of L_ω, we rewrite equation (94) in the form

$$|(I - \omega L)^{-1}\{I + \omega(U - I)\} - \lambda I| = 0,$$

$$|\{I + \omega(U - I)\} - \lambda(I - \omega L)| = 0,$$

$$\left|(U + \lambda L) - \frac{\lambda + \omega - 1}{\omega} I\right| = 0,$$

$$\left|\lambda^{\frac{1}{2}}(\lambda^{\frac{1}{2}}L + \lambda^{-\frac{1}{2}}U) - \frac{\lambda + \omega - 1}{\omega} I\right| = 0,$$

$$\left|(\lambda^{\frac{1}{2}}L + \lambda^{-\frac{1}{2}}U) - \frac{\lambda + \omega - 1}{\lambda^{\frac{1}{2}}\omega} I\right| = 0.$$

If I – L – U is two-cyclic and consistently ordered,

$$\left| (L+U) - \frac{\lambda + \omega - 1}{\lambda^{\frac{1}{2}}\omega} I \right| = 0,$$

and so for any eigenvalue λ of the S.O.R. matrix L_ω, there corresponds an eigenvalue μ of the point Jacobi matrix $(L + U)$, where

$$\mu = \frac{\lambda + \omega - 1}{\omega\lambda^{\frac{1}{2}}} \tag{95}$$

Thus equation (95) connects the eigenvalues of the S.O.R. matrix with the eigenvalues of the point Jacobi matrix, provided that $I - L - U$ is two-cyclic and consistently ordered.

Exercise

15. Using equation (95), show that for a two-cyclic consistently ordered matrix, the Gauss-Seidel method converges if the point Jacobi method converges, and does so at a faster rate.

For a full discussion of consistently ordered cyclic matrices and their properties, see chapter 4 of Varga (1962) and chapter 5 of Young (1971).

We now assume that $I - L - U$ is *symmetric* as well as being two-cyclic and consistently ordered, and so $(L + U)$ is symmetric and hence the eigenvalues of $(L + U)$ are *real*. It is now shown that since $(L + U)$ is weakly two-cyclic, its non-zero eigenvalues occur in $\pm$ pairs, and so $-\rho(L + U) \leqslant \mu \leqslant \rho(L + U)$.

After interchanging rows and corresponding columns, $(L + U)$ can be written as

$$\begin{bmatrix} 0_1 & F \\ G & 0_2 \end{bmatrix},$$

where 0_1 and 0_2 are null square matrices of order r and s respectively and $(L + U)$ is square of order $(r + s)$. Since the interchange of rows and corresponding columns does not affect the eigenvalues of a matrix, the eigenvalues of $(L + U)$ are given by

$$\begin{vmatrix} -\mu I_1 & F \\ G & -\mu I_2 \end{vmatrix} = 0$$

where I_1 and I_2 are unit matrices of order r and s respectively. Multiplying the first r rows and the last s columns of the determinant by -1, the result

$$\begin{vmatrix} \mu I_1 & F \\ G & \mu I_2 \end{vmatrix} = 0$$

is obtained, showing that $-\mu$ is also an eigenvalue of $(L + U)$.

It is now assumed that the point Jacobi method is convergent and so $0 < \rho(L + U) < 1$. In addition, since $I - L - U$ is a consistently ordered two-cyclic matrix, the Gauss-Seidel method (S.O.R. with $\omega = 1$) is also convergent (see Exercise 15), and so we hope that the S.O.R. method is convergent for a range of ω including $\omega = 1$.

Returning to equation (95), we now consider, for a given value of μ in the range $0 < \mu \leqslant \rho(L + U) < 1$, the two functions of λ,

$$f_\omega(\lambda) = \frac{\lambda + \omega - 1}{\omega}$$

and

$$g(\lambda) = \mu \lambda^{\frac{1}{2}}.$$

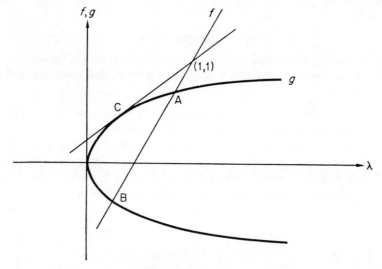

Figure 8

These functions are shown in figure 8, where $f_\omega(\lambda)$ is a straight line passing through $(1, 1)$ and $g(\lambda)$ is a parabola. Result (95) can now be interpreted geometrically as the intersection of the curves $f_\omega(\lambda)$ and $g(\lambda)$, with the two values of λ at the points of intersection (A and B) given by

$$\lambda^2 + 2[(\omega - 1) - \tfrac{1}{2}\mu^2 \omega^2] \lambda + (\omega - 1)^2 = 0,$$

and so

$$\lambda = \tfrac{1}{2}\mu^2 \omega^2 - (\omega - 1) \pm \mu\omega[\tfrac{1}{4}\mu^2 \omega^2 - (\omega - 1)]^{\frac{1}{2}}.$$

The larger abscissa of the two points of intersection decreases with increasing ω, until eventually $f_\omega(\lambda)$ becomes a tangent to $g(\lambda)$ at the point C. In this case

$$\tfrac{1}{4}\mu^2 \omega^2 - \omega + 1 = 0,$$

from which

$$\omega = \frac{1 \pm (1 - \mu^2)^{\frac{1}{2}}}{\frac{1}{2}\mu^2}$$

$$= \frac{2}{1 \mp (1 - \mu^2)^{\frac{1}{2}}} .$$

With −ve sign, $0 \leqslant \mu \leqslant 1$ gives $\infty \geqslant \omega \geqslant 2$.
With +ve sign, $0 \leqslant \mu \leqslant 1$ gives $1 \leqslant \omega \leqslant 2$.
The range of ω must include $\omega = 1$ and so we choose

$$\bar{\omega} = \frac{2}{1 + (1 - \mu^2)^{\frac{1}{2}}} . \tag{96}$$

If $\omega > \bar{\omega}$, λ has two complex conjugate roots, viz.

$$\lambda = \frac{1}{2}\mu^2 \omega^2 - (\omega - 1) \pm i\mu\omega \left[(\omega - 1) - \frac{1}{4}\mu^2 \omega^2 \right]^{\frac{1}{2}},$$

from which it follows that

$$|\lambda| = \omega - 1.$$

Thus the minimum value of λ is

$$\tilde{\lambda} = \bar{\omega} - 1,$$

where $\bar{\omega}$ is given by equation (96) and μ is an eigenvalue of $(L + U)$ in the range $0 < \mu \leqslant \rho(L + U) < 1$. Since

$$g(\lambda) = \rho(L + U)\lambda^{\frac{1}{2}}$$

is the envelope of all the curves

$$g(\lambda) = \mu\lambda^{\frac{1}{2}},$$

where $0 < \mu \leqslant \rho(L + U) < 1$, it follows that

$$\min_{\omega} \rho(L_\omega) = \rho(L_{\omega_0}) = \omega_0 - 1, \tag{97}$$

where ω_0 is given by

$$\omega_0 = \frac{2}{1 + (1 - \mu_0^2)^{\frac{1}{2}}} \tag{98}$$

where

$$\mu_0 = \rho(L + U).$$

We have thus found the value of ω, given by equation (98), which minimizes the maximum modulus eigenvalue of L_ω.

So far μ has been considered in the range $0 < \mu \leqslant \rho(L + U) < 1$. If $-1 < \rho(L + U) \leqslant \mu < 0$, then

$$g(\lambda) = -\mu\lambda^{\frac{1}{2}},$$

and the arguments go through as before. Finally, it should be pointed out that the point Jacobi method is convergent if $0 < \rho(L + U) < 1$ and so from equation (98) it follows that $1 < \omega_0 < 2$, and from equation (97) that $0 < \rho(L_{\omega_0}) < 1$. A summary will now be given of the method of S.O.R.

Summary

The theory of S.O.R. can be applied to the solution of the system of equations

$$(I - L - U)\mathbf{x} = \mathbf{b},$$

if the matrix $(I - L - U)$ is symmetric, two-cyclic and consistently ordered. The theory consists of the following steps:

(i) Calculate $\mu_0 = \rho(L + U)$ which must satisfy $0 < \mu_0 < 1$.

(ii) From equation (98) obtain the optimum overrelaxation factor ω_0 where $1 < \omega_0 < 2$.

(iii) Insert this optimum value ω_0 into equation (93) and the method of S.O.R. becomes

$$\mathbf{x}_{i+1} = (I - \omega_0 L)^{-1}[\omega_0 U + (1 - \omega_0)I]\mathbf{x}_i + \omega_0(I - \omega_0 L)^{-1}\mathbf{b}.$$

Example 9. Show that the five-point difference replacement of Laplace's equation on the unit square subject to Dirichlet boundary conditions leads to a two-cyclic coefficient matrix.

Let $x_{l,m}$ denote the component of $\mathbf{x}$ at the grid point (lh, mh). The vector $\mathbf{x}$ is partitioned so that

$$\mathbf{x} = \begin{bmatrix} \mathbf{x}_{\text{even}} \\ \mathbf{x}_{\text{odd}} \end{bmatrix}$$

where the component $x_{l,m}$ belongs to $\mathbf{x}_{\text{even}}$ if $l + m$ is even and to $\mathbf{x}_{\text{odd}}$ when $l + m$ is odd. The corresponding ordering of the components of $\mathbf{x}$ is shown in figure 9 for $M = 4$ where grid values at nodes 1–5 belong to $\mathbf{x}_{\text{even}}$ and those at nodes 6–9 belong to $\mathbf{x}_{\text{odd}}$. The difference equation at node (lh, mh) is

$$x_{l,m} = \tfrac{1}{4}(x_{l+1,m} + x_{l-1,m} + x_{l,m+1} + x_{l,m-1}) \tag{99}$$

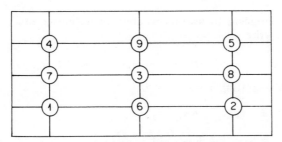

Figure 9

and when $x_{l,m}$ is a component of $\mathbf{x}_{even}$, all the terms on the right are components of $\mathbf{x}_{odd}$. The totality of equations (99) may therefore be written as

$$\mathbf{x}_{even} = F\mathbf{x}_{odd} + \mathbf{b}_{odd} \qquad (100)$$

for even values of $l + m$ and as

$$\mathbf{x}_{odd} = -G\mathbf{x}_{even} + \mathbf{b}_{even} \qquad (101)$$

for odd values of $l + m$ where $\mathbf{b}_{odd}$ and $\mathbf{b}_{even}$ arise from boundary conditions. Equations (100) and (101) combine to give

$$\begin{bmatrix} I_1 & F \\ G & I_2 \end{bmatrix} \begin{bmatrix} \mathbf{x}_{even} \\ \mathbf{x}_{odd} \end{bmatrix} = \begin{bmatrix} \mathbf{b}_{odd} \\ \mathbf{b}_{even} \end{bmatrix} \qquad (102)$$

where I_1 and I_2 are square unit matrices. We have therefore defined an ordering of the unknowns so that the coefficient matrix is two-cyclic. Any other ordering of the unknowns also leads to a two-cyclic coefficient matrix since it may be obtained by a permutation of the rows and corresponding columns of the matrix of (102).

Example 10. If A is the matrix obtained for the problem of the previous example using a natural ordering of grid points, calculate the optimum value of the relaxation parameter.

Let $h = 1/M$ denote the grid spacing, then for the natural ordering of grid points

$$\mathbf{x} = \{x_{1,1}, x_{2,1}, \ldots, x_{M-1,1}; x_{1,2}, \ldots, x_{M-1,2}; \ldots; x_{1,M-1}, \ldots, x_{M-1,M-1}\}^T$$

and the ordering is depicted in figure 10 for $M = 4$.

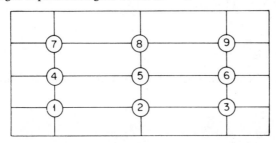

Figure 10

It is more convenient to use the notation of difference equations rather than matrices and we write

$$(A\mathbf{x})_{l,m} \equiv x_{l,m} - \tfrac{1}{4}(x_{l+1,m} + x_{l-1,m} + x_{l,m+1} + x_{l,m-1}) = 0.$$

Therefore, for the splitting $A = I - L - U$,

and
$$(L\mathbf{x})_{l,m} = \tfrac{1}{4}(x_{l-1,m} + x_{l,m-1})$$
$$(U\mathbf{x})_{l,m} = \tfrac{1}{4}(x_{l+1,m} + x_{l,m+1}).$$

It has been shown in Example 9 that A is two-cyclic and, in order to apply the necessary theory, we must now show that A is consistently ordered. For integers p, q $(0 < p, q < M)$, define the vector $\mathbf{v}$ to have components

$$v_{l,m} = \alpha^{l+m} \sin lp\pi/M \sin mq\pi/M \qquad (103)$$

for $1 \leqslant l, m \leqslant M-1$. Now using elementary trigonometric identities, it is easily shown that

$$(\alpha L\mathbf{v} + \alpha^{-1} U\mathbf{v})_{l,m} = \mu v_{l,m}$$

for all $\alpha \neq 0$, where

$$\mu \equiv \mu_{p,q} = \tfrac{1}{2}(\cos p\pi/M + \cos q\pi/M).$$

Thus, for each p, q, the vector $\mathbf{v}$ is an eigenvector of the matrix $\alpha L + \alpha^{-1} U$ with corresponding eigenvalue $\mu_{p,q}$ which is independent of α. A is therefore consistently ordered. We may also deduce that $\mu_{p,q} (1 \leqslant p, q \leqslant M-1)$ are the eigenvalues of the Jacobi matrix $L + U$ and consequently

$$\rho(L + U) \equiv \max_{p,q} |\mu_{p,q}| = \cos \pi h.$$

Hence $\mu_0 = \cos \pi h$ and, from (98), the optimum relaxation parameter is given by

$$\omega_0 = 2/(1 + \sin \pi h).$$

In a popular extension of S.O.R., called the *line* S.O.R. method, the variables on an entire line of the grid are updated simultaneously. The additional effort of solving a tridiagonal system of equations at each step is offset by a more rapid rate of convergence (see Varga (1962)).

3.14 A.D.I. methods

The Peaceman–Rachford method (2.71a) for solving the heat conduction equation

$$\frac{\partial u}{\partial t} = \frac{\partial^2 u}{\partial x_1^2} + \frac{\partial^2 u}{\partial x_2^2}$$

is given by

$$(1 - \tfrac{1}{2}r\delta_{x_1}^2)(1 - \tfrac{1}{2}r\delta_{x_2}^2)U_{l,m}^{n+1} = (1 + \tfrac{1}{2}r\delta_{x_1}^2)(1 + \tfrac{1}{2}r\delta_{x_2}^2)U_{l,m}^n \qquad (104)$$

after elimination of the intermediate solution. Now equation (104) can be taken to represent an iteration procedure which converges if

$$U_{l,m}^n = U_{l,m}^{n+1} = U_{l,m}$$

for n sufficiently large, and substitution of these values into equation (104) leads to

$$(\delta_{x_1}^2 + \delta_{x_2}^2)U_{l,m} = 0,$$

which is the standard *five-point* difference replacement of Laplace's equation. Thus the Peaceman–Rachford method given by (2.71a) (in chapter 2) applied to a heat conduction problem where the boundary conditions are independent of the time, represents an iterative method for solving Laplace's equation in a square with Dirichlet boundary conditions. This result is true, but the argument given here is not enough to justify it. The result given in Exercise 16 following equations (105), however, together with the proof of convergence, complete the justification. The quantity r is no longer the mesh ratio but is an *iteration parameter,* which may be varied from iteration to iteration. We now examine the convergence of the Peaceman–Rachford iteration procedure for solving Laplace's equation and determine successive values of r, called a *parameter sequence,* which will optimize the convergence of the process.

The totality of equations (2.71a) taken over the $(M - 1)^2$ internal grid points of the unit square leads to the matrix equations

$$\left. \begin{array}{l} \left(\dfrac{2}{r_{n+1}}I + H \right)\mathbf{U}^{n+1*} = \left(\dfrac{2}{r_{n+1}}I - V \right)\mathbf{U}^n + \mathbf{k} \\[3mm] \\ \left(\dfrac{2}{r_{n+1}}I + V \right)\mathbf{U}^{n+1} = \left(\dfrac{2}{r_{n+1}}I - H \right)\mathbf{U}^{n+1*} + \mathbf{k}, \end{array} \right\} \qquad (105)$$

and

where the parameter r_{n+1} can vary from iteration to iteration.

Exercise

16. Eliminate U^{n+1*} from equations (105) and show that the resulting equation reduces to

$$(H + V)\mathbf{U} = \mathbf{k}$$

when

$$\mathbf{U}^{n+1} = \mathbf{U}^n = \mathbf{U}.$$

F

In equations (105) H, V are matrices of order $(M-1)^2$ given by

$$H = \begin{bmatrix} C & & & & \mathbf{O} \\ & C & & & \\ & & \cdot & & \\ & & & \cdot & \\ \mathbf{O} & & & & C \end{bmatrix} \qquad V = \begin{bmatrix} 2J & -J & & & \mathbf{O} \\ -J & 2J & -J & & \\ & \cdot & \cdot & \cdot & \\ & & \cdot & \cdot & \cdot \\ & & -J & 2J & -J \\ \mathbf{O} & & & -J & 2J \end{bmatrix}$$

with C a matrix of order $(M-1)$ given by

$$C = \begin{bmatrix} 2 & -1 & & & \\ -1 & 2 & -1 & & \\ & \cdot & \cdot & \cdot & \\ & & \cdot & \cdot & \cdot \\ & & -1 & 2 & -1 \\ & & & -1 & 2 \end{bmatrix}$$

and J the unit matrix of order $(M-1)$. In equations (105), I is the unit matrix of order $(M-1)^2$, $\mathbf{U}$ and $\mathbf{k}$ are vectors given by

$$\{U_{1,1} \ldots U_{1,M-1}; U_{2,1} \ldots U_{2,M-1}; \ldots; U_{M-1,1} \ldots U_{M-1,M-1}\}^T$$

and

$$\{k_{1,1} \ldots k_{1,M-1}; k_{2,1} \ldots k_{2,M-1}; \ldots; k_{M-1,1} \ldots k_{M-1,M-1}\}^T$$

respectively, where the components of the vector $\mathbf{k}$ arise from the boundary values. The two equations of (105) are now combined to give

$$\mathbf{U}^{n+1} = T_{n+1} \mathbf{U}^n + \mathbf{g}_{n+1} \tag{106}$$

where

$$T_{n+1} = \left(\frac{2}{r_{n+1}} I + V \right)^{-1} \left(\frac{2}{r_{n+1}} I + H \right)^{-1} \left(\frac{2}{r_{n+1}} I - H \right) \left(\frac{2}{r_{n+1}} I - V \right) \tag{107}$$

and

$$\mathbf{g}_{n+1} = \left(\frac{2}{r_{n+1}} I + V \right)^{-1} \left\{ \left(\frac{2}{r_{n+1}} I + H \right)^{-1} \left(\frac{2}{r_{n+1}} I - H \right) + I \right\} \mathbf{k}.$$

Since

$$HV = VH,$$

the matrices possess a common set of orthonormal eigenvectors $\alpha^{p,q}$, $(1 \leqslant p,q \leqslant M - 1)$ with the corresponding eigenvalues given by

$$H\alpha^{p,q} = 4 \sin^2\left(\frac{p\pi}{2M}\right)\alpha^{p,q}$$

$$V\alpha^{p,q} = 4 \sin^2\left(\frac{q\pi}{2M}\right)\alpha^{p,q},$$

where $1 \leqslant p,q \leqslant M - 1$. Hence from equation (107)

$$T_{n+1}\alpha^{p,q} = \frac{\left(\dfrac{2}{r_{n+1}} - 4 \sin^2 \dfrac{p\pi}{2M}\right)\left(\dfrac{2}{r_{n+1}} - 4 \sin^2 \dfrac{q\pi}{2M}\right)}{\left(\dfrac{2}{r_{n+1}} + 4 \sin^2 \dfrac{p\pi}{2M}\right)\left(\dfrac{2}{r_{n+1}} + 4 \sin^2 \dfrac{q\pi}{2M}\right)}\alpha^{p,q}.$$

Thus the eigenvalues $\lambda_{p,q}(r_{n+1})$ of T_{n+1} may be conveniently expressed as

$$\lambda_{p,q}(r_{n+1}) = \frac{(1-a)(1-b)}{(1+a)(1+b)} \tag{108}$$

where

$$a = 2r_{n+1} \sin^2 \frac{p\pi}{2M} \quad \text{and} \quad b = 2r_{n+1} \sin^2 \frac{q\pi}{2M}$$

with $1 \leqslant p,q \leqslant M - 1$. Since $a > 0$, $b > 0$, it can easily be shown that

$$-1 < \lambda_{p,q}(r_{n+1}) < 1$$

and consequently

$$\rho(T_{n+1}) < 1.$$

If r_{n+1} is *constant for all n*, then T_{n+1} is also constant, and the iterative process given by equation (106) is convergent since $\rho(T_{n+1}) < 1$. This follows from the discussion of the general iterative process (90). This leads to the important result that the Peaceman–Rachford iterative method for solving Laplace's equation in a square is convergent for any *fixed* value of the iteration parameter. The optimum value of the fixed iteration parameter $r_{n+1} \equiv r$ may be deduced by minimizing the spectral radius $\rho(T_{n+1})$. To this end we note that

$$\max_{\mu \leqslant a,\, b \leqslant v} \left| \frac{(1-a)(1-b)}{(1+a)(1+b)} \right| = \max \left\{ \left(\frac{1-\mu}{1+\mu} \right)^2, \left(\frac{1-v}{1+v} \right)^2 \right\} \qquad (109)$$

where, in view of the definitions of a and b, $\mu = 2r \sin^2 \pi/2M$ and $v = 2r \cos^2 \pi/2M$. The right hand side of (109) is minimized when $\mu v = 1$ from which we find

$$r = \operatorname{cosec} \pi/M$$

as the optimal value of the fixed iteration parameter. It can further be shown that the spectral radius (and, consequently, the rate of convergence) of this method is asymptotically the same as S.O.R. with optimum relaxation parameter.

If the iteration parameter r_{n+1} in the equations of (105) is allowed to vary and take the values r_{n+1} $[n = 0, 1, \ldots, (n_0 - 1)]$ for each of n_0 $(\geqslant 2)$ successive iterations, a substantial improvement in the convergence of the Peaceman–Rachford method for solving Laplace's equation in a square can be obtained. A full discussion of optimum parameters for the model problem — the Dirichlet problem for a unit square — is well beyond the scope of this book. The interested reader is referred to Varga (1962), Wachspress (1966) and Young (1971).

The difficulty of determining optimal parameters has led a number of authors to devise alternative sequences. An example is given by Wachspress (1957) in which

$$r_{n+1} = \frac{1}{2 \cos^2 (\pi/2M)} \left(\cot^2 \frac{\pi}{2M} \right)^{n/(n_0 - 1)} \qquad (n = 0, 1, \ldots, n_0 - 1),$$

where n_0 $(\geqslant 2)$ is the *smallest* integer such that

$$(\sqrt{2} - 1)^{(n_0 - 1)} \leqslant \tan \pi/2M.$$

The sequence is applied cyclically until convergence is attained.

Example 11. Solve Laplace's equation in the unit square subject to the boundary conditions

$$u(0, x_2) = u(1, x_2) = 0 \quad (0 \leqslant x_2 \leqslant 1);$$
$$u(x_1, 0) = u(x_1, 1) = \sin \pi x_1 \quad (0 \leqslant x_1 \leqslant 1),$$

using the Peaceman–Rachford method with Wachspress iteration parameters.

The calculation is carried out on a digital computer for the case of 100 internal grid points, that is, $M = 11$. The condition

$$(\sqrt{2} - 1)^{(n_0 - 1)} \leqslant \tan \frac{\pi}{22}$$

leads to $n_0 = 4$ and the parameter sequence is given by

0.51033610,
1.85949096,
6.77535187,
24.68707504.

The theoretical solution of the problem is

$$u(x_1, x_2) = \text{sech } \tfrac{1}{2}\pi \cosh \pi(x_2 - \tfrac{1}{2}) \sin \pi x_1.$$

The iterative procedure starts with $u(x_1, x_2) = 0$ for all grid points (x_1, x_2) inside the unit square. The error at the centre grid point of the square after four iterations, that is, using the parameter sequence once, is

$$- 0.00382376.$$

Parameter sequences can of course be obtained for A.D.I. methods, other than the Peaceman—Rachford. A technique for solving difference replacements of Laplace's equation in regions with boundaries parallel to the co-ordinate axes is to combine A.D.I. methods with the *Schwarz alternating procedure*. The Schwarz procedure, which is described in Kantorovich and Krylov (1958), enables one to solve the Dirichlet problem for Laplace's equation on the union of two overlapping plane regions, provided the Dirichlet problem is solvable on each separately, and provided the boundaries of the regions intersect at non-zero angles.

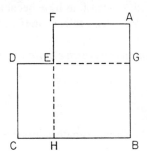

Figure 11

To illustrate this alternating procedure, we consider the solution of Laplace's equation in the L-shaped region shown in figure 11. This region consists of two overlapping rectangles, BCDG (region R_1) and ABHF (region R_2). The boundary conditions consist of u given on the perimeter of the L-shaped region. The region is covered by a grid with lines parallel to the sides, and initially we put $U^0_{l,m} = 0$ at all grid points lying inside the region. The first calculation consists of using an A.D.I. procedure together with the appropriate parameter sequence to solve the Dirichlet problem in the region R_1, the values of U being prescribed as zero at grid points on EG. This calculation gives a first estimate of the values of U at grid points along EH, which enables a solution to be obtained in R_2, again using the equations (2.71a) together with the parameter sequence and dropping the originally prescribed values on EG. A new estimate of the value of U along EG is thus obtained.

This procedure, which is convergent, is continued until a value of n is reached, such that

$$| U_{l,m}^{n+1} - U_{l,m}^n | < \delta,$$

for all grid points inside the L-shaped region, where δ is a measure of the accuracy required. In principle, there is no limit to the number of overlapping regions to which the Schwarz alternating procedure can be extended. In practice, however, the method becomes rather tedious if a large number of overlapping regions occur.

It is worth pointing out that in many physical problems, the boundary conditions in the neighbourhood of the re-entrant corner lead to singularities in some of the derivatives of the solution at the corner. This, and similar situations, are discussed more fully in Chapter 6.

3.15 Conjugate gradient and related methods

Although the conjugate gradient method was devised as long ago as 1952 by Hestenes and Stiefel, it is only recently that it has gained acceptance as a method for solving finite difference equations. It is still only rarely used as a method in its own right, rather it is used as a means of accelerating otherwise slowly convergent iterative processes. The necessary extensions have become known as preconditioned conjugate gradient methods. We begin with a brief account of the standard conjugate gradient algorithm.

Let A denote an $N \times N$ *symmetric positive definite* matrix, then we can associate with the algebraic system

$$Ax = b \tag{110}$$

a measure of the error $E(y)$ of an approximate solution y given by

$$E(y) = \tfrac{1}{2}(y - x)^T A (y - x) \tag{111}$$

where x is the solution of (110). By virtue of the positive definiteness of A, $E(y)$ has a unique minimum at the point $y = x$. Equation (111) may be written in several alternative forms, including

$$E(y) = \tfrac{1}{2}y^T A y - b^T y$$

$$= \tfrac{1}{2}r^T A^{-1} r$$

where $r = b - Ay$ is called the *residual vector*.

To describe the algorithm in general terms, we assume that at the kth step we have available the current approximation x_k, the associated residual $r_k = b - Ax_k$

and a search direction p_k. The next approximation x_{k+1} is determined by minimizing E in the direction p_k, that is, α_k is chosen to minimize the expression

$$E(x_k + \alpha_k p_k)$$

and we then set $x_{k+1} = x_k + \alpha_k p_k$. The direction of the steepest descent from x_{k+1} is given by $r_{k+1} (= b - Ax_{k+1})$, but choosing p_{k+1} to be parallel to r_{k+1} has been found not to work well in practice. However, by choosing the next search direction p_{k+1} to be the component of r_{k+1} which is conjugate to the directions $p_0, p_1, \ldots$ p_k, that is

$$p_{k+1}^T A p_j = o \quad \text{for} \quad j = 0, 1, \ldots, k,$$

the exact solution can be attained in N cycles. The success of the method depends on the fact that these search directions can be computed recursively.

The complete algorithm can be conveniently broken down into the following steps:

(i) Let $k = 0, r_0 = b - Ax_0$ and $p_0 = r_0$.
For $k = 0, 1, 2, \ldots$. compute the vectors x_k, r_k and p_k from
(ii) $x_{k+1} = x_k + \alpha_k p_k$ where $\alpha_k = r_k^T r_k / p_k^T A p_k$
(iii) $r_{k+1} = r_k - \alpha_k A p_k$
(iv) $p_{k+1} = r_{k+1} + \beta_k p_k$ where $\beta_k = r_{k+1}^T r_{k+1} / r_k^T r_k$.

Exercise

17. Prove inductively that, for fixed $k > 0$,

$$r_{k+1}^T r_j = 0$$

and

$$p_{k+1}^T A p_j = 0 \quad (j = 0, 1, 2, \ldots, k).$$

It must be emphasized, that when calculations are performed in exact arithmetic, the exact solution x of (110) is obtained in not more than N cycles (N being the dimension of x) and the method should therefore be classified as being of direct type. However, when solving large systems and rounding errors are present in the calculation, the mutual conjugacy of the sequence $p_0, p_1, \ldots$ becomes degraded and, consequently, so too is the finite termination property. Fortunately it is observed that $x_k \approx x$ for $k \ll N$, when N is large, and the algorithm can be employed in an iterative fashion.

The efficient implementation of this algorithm requires the storage of only the non-zero elements of A together with the four vectors x_k, r_k, p_k and $A p_k$. This is greater than the storage requirements of S.O.R. for instance, but may be offset by the fact that the parameters α_k, β_k are determined automatically, unlike S.O.R. where the optimum relaxation parameter may have to be estimated.

Example 12. Use the conjugate gradient method to solve the system $A\mathbf{x} = \mathbf{b}$ *where*

$$A = \begin{bmatrix} 1 & -\frac{1}{2} & 0 \\ -\frac{1}{2} & 1 & -\frac{1}{2} \\ 0 & -\frac{1}{2} & 1 \end{bmatrix}, \qquad \mathbf{b} = \begin{bmatrix} 0 \\ 0 \\ 2 \end{bmatrix}$$

with an initial approximation $\mathbf{x}_0 = [2, -3, 2]^T$.

Employing the relations of step (i) we have

$$\mathbf{r}_0 = \mathbf{p}_0 = \begin{bmatrix} -3.5 \\ 5.0 \\ -1.5 \end{bmatrix}, \qquad A\mathbf{p}_0 = \begin{bmatrix} -6.0 \\ 7.5 \\ -4.0 \end{bmatrix}$$

giving $\mathbf{r}_0^T \mathbf{r}_0 = 39.5$, $\mathbf{p}_0^T A \mathbf{p}_0 = 64.5$ and so $\alpha_0 = 0.6124$. Then for $k = 0$,

$$\mathbf{x}_1 = \begin{bmatrix} 2 \\ -3 \\ 2 \end{bmatrix} + \alpha_0 \begin{bmatrix} -3.5 \\ 5.0 \\ -1.5 \end{bmatrix} = \begin{bmatrix} -0.1431 \\ 0.0620 \\ 1.0814 \end{bmatrix} \quad \text{and} \quad \mathbf{r}_1 = \begin{bmatrix} 0.1744 \\ 0.4070 \\ 0.9500 \end{bmatrix}.$$

Therefore $\beta_0 = \mathbf{r}_1^T \mathbf{r}_1 / \mathbf{r}_0^T \mathbf{r}_0 = 1.0978/39.5 = 0.02779$.
The next search direction is

$$\mathbf{p}_1 = \mathbf{r}_1 + \beta_0 \mathbf{p}_0 = \begin{bmatrix} 0.0771 \\ 0.5459 \\ 0.9079 \end{bmatrix} \quad \text{and} \quad A\mathbf{p}_1 = \begin{bmatrix} -0.1958 \\ 0.0534 \\ 0.6350 \end{bmatrix}.$$

This completes the cycle for $k = 0$. For $k = 1$

$$\mathbf{x}_2 = \begin{bmatrix} 0.0 \\ 1.0769 \\ 2.7692 \end{bmatrix}, \qquad \mathbf{r}_2 = \begin{bmatrix} 0.5385 \\ 0.3077 \\ -0.2308 \end{bmatrix}$$

giving $\beta_1 = 0.3989$ and

$$\mathbf{p}_2 = \begin{bmatrix} 0.5692 \\ 0.5254 \\ 0.1314 \end{bmatrix}, \qquad A\mathbf{p}_2 = \begin{bmatrix} 0.3065 \\ 0.1751 \\ -0.1314 \end{bmatrix}, \qquad \alpha_2 = 1.7568.$$

Incidentally, an easy computation at this point shows that $\mathbf{p}_2^T A \mathbf{p}_1 = 0$ to four decimal places (see Exercise 17). Finally, for $k = 2$,

$$\mathbf{x}_3 = \mathbf{x}_2 + \alpha_2 \mathbf{p}_2 = \begin{bmatrix} 1.0000 \\ 2.0000 \\ 3.0000 \end{bmatrix}, \qquad \text{and} \qquad \mathbf{r}_3 = \mathbf{b} - A\mathbf{x}_3 = \mathbf{0}.$$

Therefore, as predicted in the earlier discussion, the method gives the exact solution in three cycles. This example should serve to illustrate the fact that the work involved per iteration in the conjugate gradient method is considerably greater than that for the iterative methods of section 3.13.

It has been pointed out by Reid (1972) that when the coefficient matrix is structured so that

$$A = \begin{bmatrix} I_1 & F \\ F^T & I_2 \end{bmatrix}$$

where I_1 and I_2 are unit matrices (see the discussion of two-cyclic and consistently ordered matrices in section 3.13) then the work per cycle can be approximately halved. This approach turns out to be more efficient than optimized S.O.R. for solving the Dirichlet problem in a square.

There is much evidence, both theoretical and numerical, to show that the conjugate gradient method converges particularly rapidly when a large proportion of the eigenvalues of A are clustered around a relatively small number of points on the real line. This is due in broad terms to the ability of the method to extract and utilize information regarding the distribution of eigenvalues (in the spirit of A.D.I. with an optimal parameter sequence). This contrasts with S.O.R. where the optimum relaxation factor is determined solely from the extreme eigenvalues of A.

Preconditioned conjugate gradient methods

When A is symmetric and positive definite, the essence of preconditioning (or scaling) is the determination of a suitable positive definite matrix M so that the system of equations

$$A\mathbf{x} = \mathbf{b} \tag{112}$$

may be replaced by the system

$$M^{-1}A\mathbf{x} = M^{-1}\mathbf{b}, \tag{113}$$

which is more conducive to solution via conjugate gradients. Applying the standard conjugate gradient algorithm to (113) gives

   (i) Let $k = 0$, $\mathbf{r}_0 = \mathbf{b} - A\mathbf{x}_0$ and $\mathbf{p}_0 = M^{-1}\mathbf{r}_0$.

   For $k = 0, 1, 2, \ldots$ compute the vectors $\mathbf{x}_k$, $\mathbf{r}_k$ and $\mathbf{p}_k$ from

   (ii) $\mathbf{x}_{k+1} = \mathbf{x}_k + \alpha_k \mathbf{p}_k$ where $\alpha_k = \mathbf{r}_k^T M^{-1} \mathbf{r}_k / \mathbf{p}_k^T A \mathbf{p}_k$.

   (iii) $\mathbf{r}_{k+1} = \mathbf{r}_k - \alpha_k A \mathbf{p}_k$

   (iv) $\mathbf{p}_{k+1} = M^{-1}\mathbf{r}_{k+1} + \beta_k \mathbf{p}_k$ where $\beta_k = \mathbf{r}_k^T M^{-1} \mathbf{r}_k / \mathbf{r}_{k+1}^T M^{-1} \mathbf{r}_{k+1}$.

The inversion of the matrix M is obviated by computing (in stage (iii)) the additional sequence of vectors $\mathbf{z}_0, \mathbf{z}_1, \ldots$ so that

$$M\mathbf{z}_k = \mathbf{r}_k \quad (k = 0, 1, \ldots). \tag{114}$$

This necessitates the solution of a system of linear equations in each cycle.

The essential requirements for an efficient algorithm are that the matrix M should be easily constructed, the solution of (114) should be inexpensive, and the matrix M should be close to A in the sense that a large number of the eigenvalues of $M^{-1}A$ should be nearly equal.

We shall describe two strategies for constructing M which satisfy these requirements.

(a) Preconditioning by incomplete factorization

Let D and L denote diagonal and unit lower triangular matrices respectively and define

$$M = LDL^T.$$

Unlike the direct factorization methods of section 3.12, we do *not* construct L and D so that $LDL^T = A$. Instead we insist that L have non-zero entries only in those positions which correspond to non-zero elements in the lower triangle of A. The product LDL^T is then formed and the elements of L and D computed by equating only the non-zero elements of A to those of LDL^T.

To illustrate the process we consider the Dirichlet problem for a general self-adjoint elliptic equation on the unit square with a grid of size $h = 1/M$. Using the natural (row by row) ordering of grid points and assuming a five-point difference replacement, then the nth component of Ax may be written in the form

$$(Ax)_n \equiv a_{n,n}x_n + a_{n,n+1}x_{n+1} + a_{n,n+M}x_{n+M} + a_{n-1,n}x_{n-1} + a_{n-M,n}x_{n-M}$$

$$[n = 1, 2, \ldots N(=(M-1)^2)] \tag{115}$$

where $a_{j,n} = 0$ for $j \leqslant 0$ and $a_{n,j} = 0$ for $j > N$.

The structure of A is shown in figure 12a and the form of (115) is depicted in figure 12b.

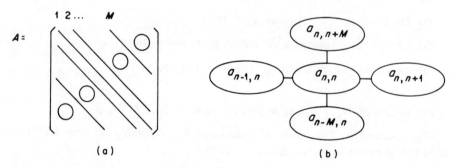

(a)   (b)

Figure 12

Suppose that D has diagonal elements $d_1, d_2, \ldots, d_N$ and L is the unit lower triangular matrix such that

$$(Lx)_n = x_n + l_{n-1,n}x_{n-1} + l_{n-M,n}x_{n-M} \quad (n = 1, 2, \ldots N), \tag{116}$$

with $l_{j,n} = 0$ for $j \leqslant 0$. The properies of L are illustrated in figure 13.

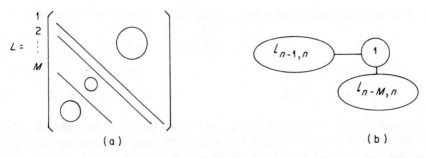

Figure 13

It follows from (116) that

$$(L^T x)_n = x_n + l_{n,n+1} x_{n+1} + l_{n,n+M} x_{n+M} \tag{117}$$

and therefore

$$(DL^T x)_n = d_n(x_n + l_{n,n+1} x_{n+1} + l_{n,n+M} x_{n+M}).$$

Also, from (116),

$$[L(DL^T x)]_n = (DL^T x)_n + l_{n-1,n}(DL^T x)_{n-1} + l_{n-M,n}(DL^T x)_{n-M}$$

which becomes

$$
\begin{aligned}
(LDL^T x)_n = {} & (d_n + d_{n-1} l_{n-1,n}^2 + d_{n-M} l_{n-M,n}^2) x_n \\
& + d_n(l_{n,n+1} x_{n+1} + l_{n,n+M} x_{n+M}) \\
& + d_{n-1} l_{n-1,n}(x_{n-1} + l_{n-1,n-M+1} x_{n+M-1}) \\
& + d_{n-M} l_{n-M,n}(x_{n-M} + l_{n-M,n-M+1} x_{n-M+1}).
\end{aligned}
\tag{118}
$$

The matrix LDL^T therefore has up to seven non-zero entries in each row (see figure 14).

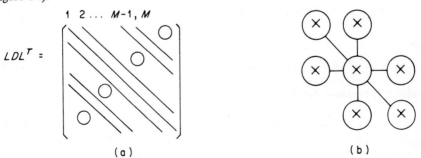

Figure 14

By equating the coefficients of x_n, x_{n+1}, x_{n+M} in (115) and (118) we obtain

$$d_n = a_{n,n} - d_{n-1}l^2_{n-1,n} - d_{n-M}l^2_{n-M,n} \qquad (n = 1, 2, \ldots N), \qquad (119)$$

$$l_{n,n+1} = a_{n,n+1}/d_n \qquad\qquad\qquad (n = 1, 2, \ldots N-1), \quad (120)$$

$$l_{n,n+M} = a_{n,n+M}/d_n \qquad\qquad\qquad (n = 1, 2, \ldots N-M), \quad (121)$$

in which $d_j = l_{j,n} = 0$ for $j \leqslant 0$ and $a_{n,j} = l_{n,j} = 0$ for $j > N$. Meijerink and Van der Vorst (1977) have proved that if the original difference equations are diagonally dominant and have non-negative coefficients in the sense described in section 3.8, then these equations uniquely define the matrices D and L. The approximate factorization need only be done once, before application of the conjugate gradient algorithm. The equation (114) is solved inexpensively by forward and backward substitution.

The eigenvalues of $M^{-1}A$ have been computed by Meijerink and Van der Vorst (1977) and Kershaw (1978) for a wide variety of problems and it is found that a large proportion of the eigenvalues are indeed closely grouped. This accounts for the rapid convergence rate observed in numerical experiments.

The incomplete factorization described here is closely related to the Strongly Implicit Method of Stone (1968) and the iterative method of Dupont, Kendall and Rachford (1968) although these authors do not use it in conjunction with conjugate gradients.

(b) Preconditioning by symmetric S.O.R.

Symmetric S.O.R. (abbreviated to S.S.O.R.) is a two-step iteration process. The first step is identical to S.O.R. and this is followed by another S.O.R. step in which the equations are taken in reverse order. A is now written in the form

$$A = D - L - L^T \tag{122}$$

where D contains the diagonal elements of A and L is a strictly lower triangular matrix (see section 3.13). From equation (91) and (92), the first step of S.S.O.R. is given by

$$D\bar{x}_{i+\frac{1}{2}} = Lx_{i+\frac{1}{2}} + Lx_i^T + b \tag{123}$$

with

$$x_{i+\frac{1}{2}} = \omega\bar{x}_{i+\frac{1}{2}} + (1 - \omega)x_i. \tag{124}$$

Similarly, the second step is given by

$$D\bar{x}_{i+1} = L^T x_{i+1} + Lx_{i+\frac{1}{2}} + b \tag{125}$$

with

$$x_{i+1} = \omega\bar{x}_{i+1} + (1 - \omega)x_{i+\frac{1}{2}} \tag{126}$$

The intermediate quantities $\bar{\mathbf{x}}_{i+\frac{1}{2}}$, $\mathbf{x}_{i+\frac{1}{2}}$ and $\bar{\mathbf{x}}_{i+1}$ may be eliminated to give the expression (129) shown in the following exercise.

Exercise

18. Show that
(i) equations (123) and (124) combine to give

$$(D - \omega L)(\mathbf{x}_{i+\frac{1}{2}} - \mathbf{x}_i) = \omega \mathbf{r}_i, \tag{127}$$

where $\mathbf{r}_i = \mathbf{b} - A\mathbf{x}_i$ and
(ii) equations (125) and (126) combine to give

$$(D - \omega L^T)(\mathbf{x}_{i+1} - \mathbf{x}_{i+\frac{1}{2}}) = \omega(\mathbf{b} - A\mathbf{x}_{i+\frac{1}{2}}). \tag{128}$$

Further, by eliminating $\mathbf{x}_{i+\frac{1}{2}}$ between (127) and (128), deduce that

$$(D - \omega L)D^{-1}(D - \omega L^T)(\mathbf{x}_{i+1} - \mathbf{x}_i) = \omega(2 - \omega)\mathbf{r}_i. \tag{129}$$

The use of S.S.O.R. for preconditioning means simply that one iteration of S.S.O.R. is interposed between each cycle of the standard conjugate gradient algorithm. In view of (129), the appropriate preconditioning matrix M takes the form

$$M = \frac{1}{\omega(2 - \omega)} (D - \omega L)D^{-1}(D - \omega L)^T.$$

Thus M is automatically constructed as a product of triangular factors whose structure is identical to the factors introduced in the previous preconditioning method (see Axelsson (1972) for further details).

Finally there is the problem of determining a suitable relaxation parameter ω. Fortunately, for $1 < \omega < 2$, the rate of convergence of the overall method is much less sensitive than S.O.R. to the choice of ω. Thus an estimate will suffice for the optimum parameter.

3.16 Eigenvalue problems

The general form of the eigenproblem for an elliptic operator L consists of the determination of non-zero eigenvalues λ and corresponding non-trivial eigenfunctions u satisfying

$$Lu = \lambda u \tag{130}$$

in a region R with *homogeneous* boundary conditions on ∂R. The boundary conditions may be of Dirichlet, Neumann or Robbins type. Covering R with a rectilinear grid and employing the finite difference approximations developed earlier in this chapter, equation (130) is replaced by the algebraic eigenvalue problem

$$A\mathbf{U} = \Lambda\mathbf{U} \tag{131}$$

where A is a square matrix with eigenvalues Λ. A discussion of the algebraic eigenvalue problem would not be appropriate here and the interested reader is referred to Wilkinson (1965) and Gourlay and Watson (1973).

A number of interesting features are illustrated by the eigenproblem for the Laplacian operator

$$\frac{\partial^2 u}{\partial x_1^2} + \frac{\partial^2 u}{\partial x_2^2} + \lambda u = 0 \tag{132}$$

in the unit square R with $u = 0$ on the boundary ∂R. It is easily verified that the functions

$$u_{p,q} = \sin p\pi x_1 \sin q\pi x_2 \quad (p,q = 1, 2, \ldots),$$

are eigenfunctions satisfying (132) with corresponding eigenvalues

$$\lambda_{p,q} = (p^2 + q^2)\pi^2 \quad (p,q = 1, 2, \ldots).$$

With the usual five-point difference replacement of (132) on a square grid of size $h(= 1/M)$ we obtain

$$U_{l-1,m} + U_{l+1,m} + U_{l,m-1} + U_{l,m+1} - (4 - \Lambda h^2)U_{l,m} = 0 \tag{133}$$

for $l,m = 1, 2, \ldots, M - 1$ and, by direct substitution, it can be shown that the eigenfunctions are

$$U_{l,m}^{(p,q)} = \sin p\pi l/M \sin q\pi m/M$$

leading to eigenvalues

$$\Lambda_{p,q} = 2(2 - \cos p\pi/M - \cos q\pi/M)/h^2 \tag{134}$$

for $p,q = 1, 2, \ldots, M - 1$.

Therefore, whereas the differential equation has an infinite number of eigenvalues, the finite difference equations provide only $(M - 1)^2$ and these will generally be approximations to the smallest eigenvalues of the operator L. This is confirmed by choosing fixed values for p and q and expanding the right hand side of (134) by Taylor series to give

$$\Lambda_{p,q} = \lambda_{p,q} + 0(h^2).$$

The first five eigenvalues (not counting multiplicities) λ and their approximations Λ are listed below for $M = 10$ where it can be seen that the accuracy of the dominant (smallest) eigenvalue is much better than for higher eigenvalues. Fortunately it is usually the dominant eigenvalue which is of greatest physical interest. It is fortuitous in this example that $U_{l,m}^{(p,q)} = u_{p,q}$ (lh, mh). In general the accuracy of the approximate eigenfunctions will be much more difficult to assess than the accuracy of the eigenvalues.

λ	Λ
19.739	19.577
49.348	47.985
78.957	76.393
128.305	120.640
177.653	164.886

Chapter 4

Hyperbolic Equations

It cannot be stressed too strongly that the study of finite difference methods of solution of hyperbolic equations is only possible after the reader has become thoroughly acquainted with the role of characteristics in the solution of hyperbolic partial differential equations. The relevant material can be obtained in works such as Garabedian (1964). Jeffrey and Tanuiti (1964), Courant and Hilbert (1962), etc.

4.1 First-order hyperbolic equations

Consider initially the almost trivial initial value problem consisting of the first-order hyperbolic equation

$$\frac{\partial u}{\partial t} + a \frac{\partial u}{\partial x} = 0 \quad (a > 0), \tag{1}$$

where a is a constant, together with the initial condition

$$u = U(x)$$

for $t = 0$ and $-\infty < x < +\infty$. There is a single family of characteristics associated with equation (1). These are the straight lines inclined to the x-axis at an angle

$$\theta = \tan^{-1}(1/a).$$

The theoretical solution of the problem is

$$u(x, t) = U(x_0),$$

where

$$x_0 = x - at.$$

The region $-\infty < x < +\infty$, $t \geq 0$ is covered by a rectangular grid with lines

164

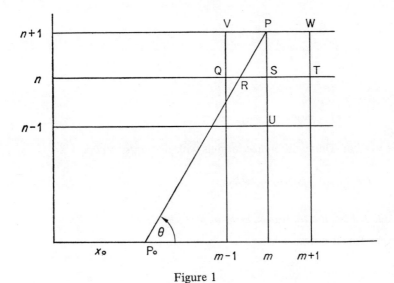

Figure 1

parallel to the x- and t-axes. The grid point P (see figure 1) is given by $x = mh$, $t = (n + 1)k$ where h and k are the grid spacings in the x- and t-directions respectively.

4.2 Explicit difference formulae

The first difference formula considered as a replacement of equation (1) involves the grid points

$$P(mh,(n + 1)k), \quad Q((m - 1)h, nk), \quad \text{and} \quad S(mh, nk).$$

Using Taylor's theorem,

$$u_P = \exp\ k\left(\frac{\partial}{\partial t}\right) u_S$$

$$= \exp\left(- ka \frac{\partial}{\partial x}\right) u_S \quad \text{(from (1))}$$

but

$$\frac{\partial}{\partial x} u_S \approx \frac{1}{h}\ (u_S - u_Q) = \frac{1}{h}\ \nabla_x u_S,$$

since

$$\nabla_x u_S = u_S - u_Q,$$

and so

$$u_P \approx (1 - pa\nabla_x)u_S$$

$$= u_S - pa(u_S - u_Q)$$

$$= (1 - pa)u_S + pau_Q,$$

where $p = k/h$. Thus a difference formula with *first-order accuracy* is

$$U_P = (1 - pa)U_S + paU_Q. \tag{2}$$

Example 1 Find the truncation error of formula (2).

From the Taylor series,

$$u_P = \left(u + k\frac{\partial u}{\partial t} + \tfrac{1}{2}k^2 \frac{\partial^2 u}{\partial t^2} + \cdots \right)_S$$

$$u_Q = \left(u - h\frac{\partial u}{\partial x} + \tfrac{1}{2}h^2 \frac{\partial^2 u}{\partial x^2} - \cdots \right)_S .$$

Substitute in (2) to give

$$\left(u + k\frac{\partial u}{\partial t} + \tfrac{1}{2}k^2 \frac{\partial^2 u}{\partial t^2} \cdots \right)_S = \left[u - pau + pa\left(u - h\frac{\partial u}{\partial x} + \tfrac{1}{2}h^2 \frac{\partial^2 u}{\partial x^2} \cdots \right) \right]_S ,$$

and so the trucation error is

$$\tfrac{1}{2}k^2 \left(\frac{\partial^2 u}{\partial t^2} - pa \frac{\partial^2 u}{\partial x^2} \right)_S + \cdots$$

If we now introduce the additional grid point $T((m + 1)h, k)$, and put

$$\frac{\partial}{\partial x} u_S \approx \frac{1}{h} \delta_x u_S,$$

where δ_x is the standard central difference operator, an expansion up to second-order terms gives

$$u_P \approx (1 - pa\delta_x + \tfrac{1}{2}p^2 a^2 \delta_x^2)u_S$$

$$= u_S - \tfrac{1}{2}pa(u_T - u_Q) + \tfrac{1}{2}p^2 a^2 (u_T - 2u_S + u_Q),$$

where the values of u at the points mid-way between Q and S, and S and T, which appear in $\delta_x u_S$ have been replaced by $\tfrac{1}{2}(u_Q + u_S)$ and $\tfrac{1}{2}(u_S + u_T)$ respectively.

Thus a difference formula of *second-order accuracy* is

$$U_P = (1 - p^2 a^2)U_S - \tfrac{1}{2}pa(1 - pa)U_T + \tfrac{1}{2}pa(1 + pa)U_Q. \tag{3}$$

This is the *Lax–Wendroff* formula, which is probably the most well known formula for first-order hyperbolic equations.

Exercise

1. Find an explicit formula of second-order accuracy involving the grid points P,S,Q, and the additional grid point

$$Z((m - 2)h, nk).$$

We now introduce the *Courant–Friedrichs–Lewy (C.F.L.) condition*, (1928), which applies to explicit difference replacements of hyperbolic equations. It requires the characteristic through P (figure 1) to intersect the line $t = nk$ within the range of points considered by the formula at the level $t = nk$. For example, R must lie between Q and S for formula (2), and between Q and T for formula (3). If the C.F.L. condition is not satisfied, convergence cannot take place when $h \to 0$, with $k = ph$ where p is constant. For formula (2)

$$\tan \theta = \frac{1}{a} = \frac{k}{RS},$$

and so

$$RS = ka.$$

The C.F.L. condition requires $0 \leqslant RS \leqslant h$, or $0 \leqslant ka \leqslant h$, or $0 \leqslant pa \leqslant 1$ ($h \neq 0$). This is also the condition for formula (3).

It is instructive to examine the stability of the Lax–Wendroff formula. Using the von Neumann method, the amplification factor ξ is easily shown to be

$$\xi = (1 - r^2) + \tfrac{1}{2}r^2(e^{i\beta h} + e^{-i\beta h}) - \tfrac{1}{2}r(e^{i\beta h} - e^{-i\beta h})$$

$$= \left(1 - 2r^2 \sin^2 \frac{\beta h}{2}\right) - ir \sin \beta h,$$

where $r = pa$. This gives

$$|\xi| = \left[1 - 4r^2(1 - r^2)\sin^4 \frac{\beta h}{2}\right]^{1/2},$$

from which it follows that

$$|\xi| \leqslant 1 \text{ if } 0 < r \leqslant 1.$$

Thus the stability condition coincides with the C.F.L. condition for the Lax–Wendroff method.

A final explicit scheme which involves three time levels and has second-order accuracy should be mentioned. It is the leap frog scheme which involves the four grid points

$$P(mh,(n + 1)k), \qquad U(mh,(n - 1)k),$$

$$Q((m - 1)h, nk) \text{ and } T((m + 1)h, nk),$$

and is given by

$$U_P = U_U - pa(U_T - U_Q). \tag{4}$$

This scheme is stable for $0 < p|a| \leqslant 1$, which coincides with the C.F.L. condition.

Example 2   Consider the initial value problem consisting of

$$\frac{\partial u}{\partial t} + \frac{\partial u}{\partial x} = 0$$

together with the initial condition

$$u = \begin{cases} 0 & (x < 0) \\ x & (0 \leqslant x \leqslant 3) \\ 6 - x & (3 \leqslant x \leqslant 6) \\ 0 & (x > 6). \end{cases}$$

Solve this problem numerically using

   (i) the Lax–Wendroff formula (3);
   (ii) the leap frog scheme (4),

and compare the numerical results with the theoretical solution, which can be used to provide extra starting values at $t = k$ for (ii).

The calculations were carried out for $h = \frac{1}{2}$ with $p = \frac{1}{2}$, and the results in table 1 were obtained after 10 time steps, i.e. at $t = 2.5$.

4.3 Implicit difference schemes

Implicit difference formulae apply only to initial boundary value problems. A typical problem of this type for a first-order hyperbolic equation consists of equation (1) together with the initial condition

$$u = U(x)$$

Table 1

x	Method (i)	Method (ii)	Theoretical solution
− 2	0	0.006	0
− 1	− 0.002	− 0.032	0
0	0.020	0.014	0
1	− 0.045	− 0.067	0
2	− 0.068	− 0.022	0
3	0.502	0.490	0.5
4	1.591	1.620	1.5
5	2.651	2.592	2.5
6	2.433	2.413	2.5
7	1.452	1.436	1.5
8	0.421	0.440	0.5
9	0.043	0.066	0
10	0.001	0.004	0
11	0	0	0
12	0	0	0

for $t = 0, 0 \leqslant x < \infty$, and the boundary condition

$$u = V(t)$$

for $x = 0, 0 \leqslant t < \infty$, where

$$U(0) = V(0).$$

The simplest implicit formula involves the grid points

$$P(mh,(n + 1)k), \quad V((m − 1)h,(n + 1)k) \text{ and } S(mh, nk)$$

(see figure 1). Using Talyor's theorem,

$$u_S = \exp\left(-k \frac{\partial}{\partial t} \right) u_P$$

$$= \exp\left(ak \frac{\partial}{\partial x} \right) u_P \quad \text{(from (1))},$$

but

$$\frac{\partial}{\partial x} u_P \approx \frac{1}{h}(u_P - u_V) = \frac{1}{h} \nabla_x u_P,$$

and so

$$u_S \approx (1 + pa\nabla_x)u_P$$

$$= u_P + pa(u_P - u_V)$$

$$= (1 + pa)u_P - pau_V.$$

The implicit difference formula is thus

$$U_S = (1 + pa)U_P - paU_V, \tag{5}$$

which is easily shown to be unconditionally stable by the method of von Neumann. Although formula (5) is an implicit formula, the method of solution is based on the explicit algorithm

$$U_m^{n+1} = \frac{pa}{1 + pa} U_{m-1}^{n+1} + \frac{1}{1 + pa} U_m^n,$$

which, together with the initial and boundary conditions, enables U to be calculated at all grid points in the quarter plane $0 < x, t < \infty$.

Exercise

2. Show that the implicit formula

$$U_S = (1 - pa)U_P + paU_W$$

is always stable.

Another implicit formula is attributed to Wendroff (1960). This involves the grid points

$$P(mh,(n + 1)k), \quad W((m + 1)h,(n + 1)k),$$

$$S(mh, nk) \quad \text{and} \quad T((m + 1)h, nk)$$

and can be obtained in the following manner:

$$\Delta_x u_S = u_T - u_S \quad \text{(by definition)},$$

and so

$$u_T = (1 + \Delta_x)u_S.$$

But

$$u_T = \left(u + h\frac{\partial u}{\partial x} + \frac{1}{2}h^2 \frac{\partial^2 u}{\partial x^2} + \dots\right)_S = \exp\left(h\frac{\partial}{\partial x}\right)u_S,$$

and so

$$1 + \Delta_x \equiv \exp\left(h \frac{\partial}{\partial x}\right),$$

or

$$\frac{\partial}{\partial x} \equiv \frac{1}{h} \log (1 + \Delta_x).$$

Similarly,

$$\frac{\partial}{\partial t} \equiv \frac{1}{k} \log (1 + \Delta_t),$$

and so substituting in equation (1),

$$\log (1 + \Delta_t) u_S = - pa \log (1 + \Delta_x) u_S,$$

or

$$u_P = (1 + \Delta_x)^{-pa} u_S. \tag{6}$$

A formula involving the grid points P, W, S and T is

$$u_P = (1 + \alpha \Delta_x)^{-1} (1 + \beta \Delta_x) u_S, \tag{7}$$

where α and β are arbitrary parameters. The right-hand sides of equations (6) and (7) are now expanded and the coefficients of Δ_x and Δ_x^2 are equated. This leads to

$$\alpha = \tfrac{1}{2}(1 + pa), \qquad \beta = \tfrac{1}{2}(1 - pa),$$

and so Wendroff's formula is

$$[1 + \tfrac{1}{2}(1 + pa)\Delta_x] U_P = [1 + \tfrac{1}{2}(1 - pa)\Delta_x] U_S. \tag{8}$$

This formula is again unconditionally stable. The method of solution based on formula (8) can be explicit in nature if (8) is written in the form

$$U_W = U_S + \frac{1 - pa}{1 + pa} (U_T - U_P),$$

which together with initial conditions on $t = 0, 0 \leqslant x < \infty$, and boundary conditions on $x = 0, 0 \leqslant t < \infty$ enables U to be calculated at all grid points in the quarter plane $0 < x, t < \infty$.

4.4 First-order hyperbolic systems in one space dimension

We now consider the first-order system of equations

$$\frac{\partial u}{\partial t} + A \frac{\partial u}{\partial x} = 0, \tag{9}$$

where A is an $n \times n$ real matrix, and u is an n-component column vector. Initially A is assumed to be constant. We consider the case where A has all real eigenvalues and n linearly independent eigenvectors, so that the system is *hyperbolic*. A is not necessarily a symmetric matrix.

Example 3. Show that (9) can be rewritten as n decoupled scalar equations

$$\frac{\partial v_j}{\partial t} + \lambda_j \frac{\partial v_j}{\partial x} = 0 \qquad (j = 1, 2, \ldots, n)$$

where λ_j $(j = 1, 2, \ldots, n)$ are the real eigenvalues of A.

A non-singular matrix P exists such that

$$PAP^{-1} = D,$$

where D is a diagonal matrix having the real eigenvalues of A as elements. Premultiplying (9) by P we get

$$\frac{\partial}{\partial t} (Pu) + PAP^{-1} \frac{\partial}{\partial x} (Pu) = 0,$$

and so

$$\frac{\partial v}{\partial t} + D \frac{\partial v}{\partial x} = 0, \tag{10}$$

where $v = Pu$. The result follows.

Example 4. In an initial boundary value problem involving boundaries at $x = 0$ and $x = 1$, determine the distribution of boundary conditions required at $x = 0, 1$ for a well-posed problem involving (9) in the case where A has k positive and $(n - k)$ negative eigenvalues.

In this problem the characteristic directions are given by

$$\frac{dx}{dt} = \lambda_j \qquad (j = 1, 2, \ldots, n),$$

where $\lambda_j > 0, j = 1, 2, \ldots, k$ and $\lambda_j < 0, j = k + 1, \ldots, n$. Each component $v_j, j = 1, 2, \ldots, n$ is obtained by integrating along the corresponding characteristic $dx/dt = \lambda_j, j = 1, 2, \ldots, n$. Consequently $v_j, j = 1, 2, \ldots, k$ can be obtained at $x = 1$ and $v_j, j = k + 1, \ldots, n$ at $x = 0$, Thus we require $(n - k)$ scalar boundary conditions at $x = 1$ and k scalar boundary conditions at $x = 0$.

N.B. If $\lambda_j = 0$, the corresponding characteristic direction is parallel to the boundaries at $x = 0, 1$, and so the component v_j can be ignored as far as boundary conditions are concerned.

Example 5. Rewrite the wave equation $\dfrac{\partial^2 \phi}{\partial t^2} = \dfrac{\partial^2 \phi}{\partial x^2}$ *as the 2 × 2 hyperbolic system (9), and find the boundary conditions required to make the solution of (9) unique in the region $[0 \leqslant x \leqslant 1] \times [t \geqslant 0]$ if the initial condition is given for $0 \leqslant x \leqslant 1$.*

If we put

$$\mathbf{u} = (u_1, u_2)^{\mathrm{T}},$$

where

$$u_1 = \frac{\partial \phi}{\partial t} \text{ and } u_2 = \frac{\partial \phi}{\partial x},$$

the wave equation $\dfrac{\partial^2 \phi}{\partial t^2} = \dfrac{\partial^2 \phi}{\partial x^2}$ can be rewritten in the form (9) where

$$A = \begin{bmatrix} 0 & -1 \\ -1 & 0 \end{bmatrix}.$$

From Example 3, if

$$P = \begin{bmatrix} 1 & -1 \\ 1 & 1 \end{bmatrix},$$

equation (9) takes the form (10) where

$$\mathbf{v} = (v_1, v_2)^{\mathrm{T}} = (u_1 - u_2, u_1 + u_2)^{\mathrm{T}},$$

and

$$D = \begin{bmatrix} 1 & 0 \\ 0 & -1 \end{bmatrix}.$$

Hence the decoupled scalar equations for v_1 and v_2 are

$$\frac{\partial v_1}{\partial t} + \frac{\partial v_1}{\partial x} = 0$$

and

$$\frac{\partial v_2}{\partial t} - \frac{\partial v_2}{\partial x} = 0$$

respectively. From the analysis following (1), the former transports v_1 without

174

change along a straight characteristic with slope $+1$ to the x-axis until the characteristic intersects the right-hand boundary. Any boundary condition at this point on $x = 1$ is permissible provided it does not contradict (or simply reproduce) the specified value of v_1. We can choose for example to specify the value of $u_1 + \alpha u_2$ ($\alpha \neq -1$) to give a well-posed condition. The left-hand boundary $x = 0$ is treated similarly.

As far as well-posed boundary conditions are concerned nothing essential is changed if, in (9), $A = A(x, t)$. The characteristics are now no longer straight lines but the solutions of the ordinary differential equations

$$\frac{dx}{dt} = \lambda_j (x, t) \qquad \text{(see Example 3).}$$

We now look at finite difference approximations to the hyperbolic system (9). The explicit and implicit difference formulae developed in the previous sections for the case of scalar a carry over in an obvious manner when A is a constant matrix. For example, the Lax–Wendroff method is now

$$\mathbf{U}_m^{n+1} = [1 - \tfrac{1}{2}pA(\Delta_x + \nabla_x) + \tfrac{1}{2}p^2 A^2 (\Delta_x - \nabla_x)] \mathbf{U}_m^n \qquad (11)$$

and Wendroff's implicit formula is

$$[I + \tfrac{1}{2}(I + pA)\Delta_x] \mathbf{U}_m^{n+1} = [I + \tfrac{1}{2}(I - pA)\Delta_x] \mathbf{U}_m^n \qquad (12)$$

where I is the unit matrix of order n, and we now write $\mathbf{U}_m^{n+1}$ and $\mathbf{U}_m^n$ for $\mathbf{U}_p$ and $\mathbf{U}_s$ respectively. If A depends on x, or x and t, (11) and (12) require modification to maintain second-order accuracy in space and time. In the latter case they become

$$\mathbf{U}_m^{n+1} = [I - \tfrac{1}{2}pA_m^{n+1/2} (\Delta_x + \nabla_x) +$$

$$\tfrac{1}{4}p^2 (A_m^{n+1/2} \Delta_x A_m^{n+1/2} \nabla_x + A_m^{n+1/2} \nabla_x A_m^{n+1/2} \Delta_x)] \mathbf{U}_m^n \qquad (11a)$$

and

$$[I + \tfrac{1}{2}(I + pA_{m+1/2}^{n+1/2}) \Delta_x] \mathbf{U}_m^{n+1} = [I + \tfrac{1}{2}(I - pA_{m+1/2}^{n+1/2})\Delta_x] \mathbf{U}_m^n \qquad (12a)$$

respectively where $A_m^{n+1/2}$ denotes evaluation of A at $x = mh$, $t = (n + \tfrac{1}{2})k$, and $A_{m+1/2}^{n+1/2}$ denotes A evaluated at $x = (m + \tfrac{1}{2})h$, $t = (n + \tfrac{1}{2}k)$.

The Courant–Friedrichs–Lewy (C.F.L.) condition for the system (9) will be mentioned only briefly. Returning to figure 1, we now have n characteristic curves passing through the point P and when the extreme curves are traced back they intersect the x-axis at points P_1 and P_n. The solution at P will be influenced only by the initial data lying between these two points and so the interval $P_1 P_n$ is referred to as the domain of dependence of P at $t = 0$. When the system (9) is replaced by an *explicit* difference scheme, it is possible to trace back, using only the structure of the difference scheme, to the grid points at $t = 0$ which influence the numerical solution at P. This set of grid points lies in an interval which we denote by D_h. The C.F.L. condition states that the domain of dependence of the difference equation must contain that of the differential equation. That is $P_1 P_n$ must lie inside D_h.

As in the scalar case, the stability of a difference approximation to a linear hyperbolic system at a grid point in the field will mean the stability of the corresponding difference equation where the values of the coefficients have been 'frozen' to the values attained at that grid point. This *local* stability condition, usually a limitation on the size of the ratio k/h, will vary from point to point in the field, and the overall stability condition must be the largest mesh ratio which satisfies the local stability condition at *every* grid point in the region.

If a typical Fourier term

$$\mathbf{U} = \mathbf{U}_0 e^{i\beta x} \quad (i = \sqrt{-1}),$$

where $\mathbf{U}_0$ is a constant vector, is substituted into the difference equation for $\mathbf{U}_m^n$, it is found that $\mathbf{U}_m^{n+1}$ is of the same form but with $G\mathbf{U}_0$ replacing $\mathbf{U}_0$. The matrix G is called the *amplification matrix*. For example, for the Lax–Wendroff method (11), the amplification matrix is

$$G = I - \tfrac{1}{2}pA(e^{i\beta h} - e^{-i\beta h}) + \tfrac{1}{2}p^2 A^2(e^{i\beta h} + e^{-i\beta h} - 2)$$

$$= [I - p^2 A^2(1 - \cos \theta)] - ipA \sin \theta. \tag{11b}$$

where $\theta = \beta h$. For the implicit Wendroff scheme (12), the amplification matrix is given implicitly by

$$[(I - pA \tan^2 \tfrac{1}{2}\theta) + i(I + pA) \tan \tfrac{1}{2}\theta] G =$$

$$[(I + pA \tan^2 \tfrac{1}{2}\theta) + i(I - pA) \tan \tfrac{1}{2}\theta]. \tag{12b}$$

The von Neumann *necessary* condition for the stability of a system is

$$\max_j |\mu_j| \leqslant 1 \qquad (j = 1, 2, \ldots, n), \tag{13}$$

where $\mu_j (j = 1, 2, \ldots, n)$ are the eigenvalues of G. It can be shown that condition (13) is satisfied for the Lax–Wendroff amplification matrix G if

$$p |\lambda_j| \leqslant 1 \qquad (j = 1, 2, \ldots, n),$$

where $\lambda_j \ (j = 1, 2, \ldots n)$ are the eigenvalues of A. This follows because from (11b), G is a rational function of A and so has the same eigenvectors as A. Hence

$$\mu_j = 1 - p^2 \lambda_j^2 (1 - \cos \theta) - ip\lambda_j \sin \theta \ (j = 1, 2, \ldots, n))$$

176

where λ_j $(j = 1, 2, \ldots, n)$ are the eigenvalues of A. As θ varies from 0 to 2π, μ_j describes an ellipse in the complex plane. This ellipse lies inside the unit circle if

$$p|\lambda_j| \leqslant 1, \qquad j = 1, 2, \ldots, n$$

and the result follows. As in the scalar case, the condition coincides with the C.F.L. condition for this problem. The implicit Wendroff scheme is unconditionally stable and so imposes no upper limit on p.

4.5 Systems of conservation laws

So far we have considered only linear first-order systems. These are represented by equation (9), where A may depend on x and t but not on $\mathbf{u}$. We now turn our attention to *non-linear* systems of the form

$$\frac{\partial \mathbf{u}}{\partial t} + \frac{\partial \mathbf{f}}{\partial x} = 0, \tag{14}$$

where $\mathbf{f}$ *is a function of* $\mathbf{u}$. Equation (14) is said to be in conservation form.

Example 6 *Show how the equations of one-dimensional compressible flow can be written in conservation form.*

x, t	distance and time coordinates respectively.
ρ	mass per unit volume.
v	velocity in the x-direction.
$m = \rho v$	momentum per unit volume.
p	pressure.
γ	ratio of specific heats.
$e = \frac{1}{2}\rho v^2 + p/(\gamma - 1)$	energy per unit volume.

Conservation of mass, energy and momentum per unit volume gives equation (14) with

$$\mathbf{u} = \begin{bmatrix} \rho \\ e \\ m \end{bmatrix}, \quad \mathbf{f}(\mathbf{u}) = \begin{bmatrix} m \\ \dfrac{\gamma e m}{\rho} - \frac{1}{2}(\gamma - 1)\dfrac{m^3}{\rho^2} \\ (\gamma - 1)e - \frac{1}{2}(\gamma - 3)\dfrac{m^2}{\rho} \end{bmatrix}$$

Now equation (14) can be rewritten in the form

$$\frac{\partial \mathbf{u}}{\partial t} + A(\mathbf{u}) \frac{\partial \mathbf{u}}{\partial x} = 0, \tag{15}$$

where $A(\mathbf{u})$ is the Jacobian of $\mathbf{f}$ with respect to $\mathbf{u}$, i.e.

$$A(\mathbf{u}) \equiv \frac{\partial(f_1, f_2, f_3)}{\partial(\rho, e, m)} \; ,$$

where

$$f_1 = m, \; f_2 = \frac{\gamma e m}{\rho} - \tfrac{1}{2}(\gamma - 1)\frac{m^3}{\rho^2} \text{ and } f_3 = (\gamma - 1)e - \tfrac{1}{2}(\gamma - 3)\frac{m^2}{\rho} \; .$$

In this example, it is easy to show that

$$A = \begin{bmatrix} 0 & 0 & 1 \\ \gamma EM + (\gamma - 1)M^3 & \gamma M & \gamma E - \tfrac{3}{2}(\gamma - 1)M^2 \\ \tfrac{1}{2}(\gamma - 3)M^2 & (\gamma - 1) & -(\gamma - 3)M \end{bmatrix},$$

where $M = m/\rho$ and $E = e/\rho$. The eigenvalues of A are real and are given by

$$u - c, u, u + c,$$

where the speed of sound c is given by $c^2 = \gamma p / \rho$. If $u > c$ (supersonic flow), the eigenvalues are all positive and A is *positive definite*.

When the physical problem is given in conservation form (14), it is instructive but by no means necessary to rewrite it in the form (15), before difference methods of solution are considered. This is illustrated with respect to the Lax–Wendroff method, which is now considered as a means of solving equation (14). Expanding by Taylor's theorem,

$$\mathbf{u}_m^{n+1} = \exp\left(k \frac{\partial}{\partial t} \right) \mathbf{u}_m^n$$

$$\approx \left(1 + k \frac{\partial}{\partial t} + \tfrac{1}{2}k^2 \frac{\partial^2}{\partial t^2} \right) \mathbf{u}_m^n$$

$$= \left[\mathbf{u} - k \frac{\partial \mathbf{f}(\mathbf{u})}{\partial x} - \tfrac{1}{2}k^2 \frac{\partial}{\partial t}\left(\frac{\partial \mathbf{f}(\mathbf{u})}{\partial x} \right) \right]_m^n \quad \text{(using (14))}$$

$$= \left[\mathbf{u} - k \frac{\partial \mathbf{f}(\mathbf{u})}{\partial x} - \tfrac{1}{2}k^2 \frac{\partial}{\partial x}\left(\frac{\partial \mathbf{f}(\mathbf{u})}{\partial t} \right) \right]_m^n$$

$$= \left[\mathbf{u} - k \frac{\partial \mathbf{f}(\mathbf{u})}{\partial x} + \tfrac{1}{2}k^2 \left\{ \frac{\partial}{\partial x}\left(A(\mathbf{u}) \frac{\partial \mathbf{f}(\mathbf{u})}{\partial x} \right) \right\} \right]_m^n,$$

where $A(\mathbf{u})$ is the Jacobian of $\mathbf{f}$ with respect to $\mathbf{u}$. Thus the Lax–Wendroff formula in conservation form is

$$U_m^{n+1} = U_m^n - \tfrac{1}{2}p(F_{m+1}^n - F_{m-1}^n) +$$

$$\tfrac{1}{2}p^2[A_{m+1/2}^n(F_{m+1}^n - F_m^n) - A_{m-1/2}^n(F_m^n - F_{m-1}^n)] \qquad (16)$$

where $F_m^n = f(U_m^n)$ and the standard central difference replacement has been used for $\partial/\partial x[A(\mathbf{u})(\partial \mathbf{f}(\mathbf{u})/\partial x)]$. In order to avoid evaluations at points other than grid points, we can rewrite

$$A_{m+1/2}^n = \tfrac{1}{2}(A_{m+1}^n + A_m^n)$$

and

$$A_{m-1/2}^n = \tfrac{1}{2}(A_m^n + A_{m-1}^n)$$

to the same order of accuracy, and equation (16) can be rewritten in the form

$$U_m^{n+1} = U_m^n - \tfrac{1}{2}p(F_{m+1}^n - F_{m-1}^n) + \tfrac{1}{2}p^2(G_m^n - G_{m-1}^n), \qquad (17)$$

where

$$G_m^n = \tfrac{1}{2}(A_{m+1}^n + A_m^n)(F_{m+1}^n - F_m^n)$$

and

$$G_{m-1}^n = \tfrac{1}{2}(A_m^n + A_{m-1}^n)(F_m^n - F_{m-1}^n).$$

Example 7 Illustrate the conservation property of the Lax–Wendroff formula (17) by using it to solve the pure initial value problem consisting of equation (14) together with $\mathbf{u} = U_0(x)$ ($-\infty < x < +\infty$) at $t = 0$.

Consider the grid points $x = mh$ on the x-axis, where m is an integer in the range $-(M + 1) \leqslant m \leqslant (M + 1)$. Applying the Lax–Wendroff formula (17) we get

$$U_{-M}^1 = U_{-M}^0 - \tfrac{1}{2}p[F_{-(M-1)}^0 - F_{-(M+1)}^0] + \tfrac{1}{2}p^2[G_{-M}^0 - G_{-(M+1)}^0]$$

$$U_{-(M-1)}^1 = U_{-(M-1)}^0 - \tfrac{1}{2}p[F_{-(M-2)}^0 - F_{-M}^0] + \tfrac{1}{2}p^2[G_{-(M-1)}^0 - G_{-M}^0]$$

$$U_{-(M-2)}^1 = U_{-(M-2)}^0 - \tfrac{1}{2}p[F_{-(M-3)}^0 - F_{-(M-1)}^0] + \tfrac{1}{2}p^2[G_{-(M-2)}^0 - G_{-(M-1)}^0]$$

$$\vdots \qquad \vdots \qquad \vdots \qquad \vdots$$

$$U_{M-2}^1 = U_{M-2}^0 - \tfrac{1}{2}p[F_{M-1}^0 - F_{M-3}^0] + \tfrac{1}{2}p^2[G_{M-2}^0 - G_{M-3}^0]$$

$$U_{M-1}^1 = U_{M-1}^0 - \tfrac{1}{2}p[F_M^0 - F_{M-2}^0] + \tfrac{1}{2}p^2[G_{M-1}^0 - G_{M-2}^0]$$

$$U_M^1 = U_M^0 - \tfrac{1}{2}p[F_{M+1}^0 - F_{M-1}^0] + \tfrac{1}{2}p^2[G_M^0 - G_{M-1}^0].$$

Adding, the result

$$\sum_{-M \leqslant m \leqslant M} \mathbf{U}_m^1 = \sum_{-M \leqslant m \leqslant M} \mathbf{U}_m^0 -$$

$$\tfrac{1}{2}p(\mathbf{F}_{M+1}^0 + \mathbf{F}_M^0 - \mathbf{F}_{-M}^0 - \mathbf{F}_{-(M+1)}^0) + \tfrac{1}{2}p^2(\mathbf{G}_M^0 - \mathbf{G}_{-(M+1)}^0)$$

is obtained. Thus **U** is conserved at two neighbouring time levels except possibly for 'flux' through side boundaries, which can be moved infinitely far apart by letting $M \to \infty$.

We now turn to the most popular type of difference method for solving (14). It consists of writing the second order Lax–Wendroff method as a two-step process. In order to facilitate the writing of the difference formulae we list the following difference operators

$$\Delta_x U_m = U_{m+1} - U_m$$

$$\nabla_x U_m = U_m - U_{m-1}$$

$$\delta_x U_m = U_{m+1/2} - U_{m-1/2}$$

$$H_x U_m = U_{m+1} - U_{m-1}$$

$$\mu_x U_m = \tfrac{1}{2}(U_{m+1/2} + U_{m-1/2}).$$

The first scheme of this type, due to Richtmyer (1963), consists of a Lax type step followed by a leap frog step and is given by

$$\bar{\mathbf{U}}_m = \mu_x U_m^n - \tfrac{1}{2}p\delta_x \mathbf{F}_m^n,$$

$$\mathbf{U}_m^{n+1} = \mathbf{U}_m^n - p\delta_x \bar{\mathbf{F}}_m, \tag{18}$$

where only *central* differences are employed. In fact, $\bar{\mathbf{U}}_m$ is a first-order approximation to $\mathbf{U}_m^{n+1/2}$. Two alternative splittings introduced by MacCormack are

$$\bar{\mathbf{U}}_m = \mathbf{U}_m^n - p\nabla_x \mathbf{F}_m^n,$$

$$\mathbf{U}_m^{n+1} = \tfrac{1}{2}(\mathbf{U}_m^n + \bar{\mathbf{U}}_m) - \tfrac{1}{2}p\Delta_x \bar{\mathbf{F}}_m \tag{19}$$

and

$$\bar{\mathbf{U}}_m = \mathbf{U}_m^n - p\Delta_x \mathbf{F}_m^n,$$

$$\mathbf{U}_m^{n+1} = \tfrac{1}{2}(\mathbf{U}_m^n + \bar{\mathbf{U}}_m) - \tfrac{1}{2}p\nabla_x \bar{\mathbf{F}}_m. \tag{20}$$

In the former, a backward difference is followed by a forward difference (BF) whereas the latter applies the differences in reverse order (FB). If in schemes (18), (19), and (20) we put

$$\mathbf{F} = A\mathbf{U} \qquad A \text{ constant matrix}$$

at the grid points and eliminate the intermediate value $\bar{\mathbf{U}}_m$, we get, in all cases

$$\mathbf{U}^{n+1} = [I - \tfrac{1}{2}pAH_x + \tfrac{1}{2}p^2 A^2 \delta_x^2]\, \mathbf{U}^n$$

which is the Lax–Wendroff formula (11). McGuire and Morris (1973) have generalized (18) in time by introducing the parameter s $(0 < s \leqslant 1)$ to give the two-step process

$$\bar{\mathbf{U}}_m = \mu_x \mathbf{U}_m^n - sp\delta_x \mathbf{F}_m^n,$$

$$\mathbf{U}_m^{n+1} = \mathbf{U}_m^n - \tfrac{1}{2}p[(1 - \frac{1}{2s})H_x \mathbf{F}_m^n + \frac{1}{s}\,\delta_x \bar{\mathbf{F}}_m]. \tag{21}$$

This reduces to (18) when $s = \tfrac{1}{2}$ and so to a scheme given by Rubin and Burstein (1967) when $s = 1$. Again if we put $\mathbf{F} = A\mathbf{U}$ at grid points, elimination of $\bar{\mathbf{U}}_m$ leads to the Lax–Wendroff formula (11).

4.6 First-order hyperbolic systems in two space dimensions.

The material of the two previous sections is now extended to cover first-order hyperbolic systems in two or more space dimensions. In the two space dimensional case we first consider

$$\frac{\partial \mathbf{u}}{\partial t} + A\,\frac{\partial \mathbf{u}}{\partial x} + B\,\frac{\partial \mathbf{u}}{\partial y} = 0 \tag{22}$$

where A, B are real constant $n \times n$ matrices and $\mathbf{u}$ is an n-component column vector. Equation (22) represents a *hyperbolic* system if, for all real α, β with $\alpha^2 + \beta^2 = 1$, there exists a non-singular transformation matrix P such that

$$P(\alpha A + \beta B)P^{-1} = D$$

where D is a diagonal matrix with real elements. Symmetry of A and B is sufficient to guarantee that (22) is a system of hyperbolic type.

Example 8 Show that Maxwell's equations

$$\dot{\mathbf{B}} + \Delta \times \mathbf{E} = 0$$

$$\dot{\mathbf{E}} - \Delta \times \mathbf{B} = 0,$$

where the dot denotes partial differentiation with respect to the time, can be written as a first-order symmetric hyperbolic system.

If

$$\mathbf{Q} = [E_x, E_y, E_z, B_x, B_y, B_z]^T,$$

then

$$\frac{\partial \mathbf{Q}}{\partial t} + A_1 \frac{\partial \mathbf{Q}}{\partial x} + A_2 \frac{\partial \mathbf{Q}}{\partial y} + A_3 \frac{\partial \mathbf{Q}}{\partial z} = 0$$

where

$$A_1 = \begin{bmatrix} . & . & . & . & . & . \\ . & . & . & . & . & -1 \\ . & . & . & . & 1 & . \\ . & . & . & . & . & . \\ . & . & 1 & . & . & . \\ . & -1 & . & . & . & . \end{bmatrix}, \qquad A_2 = \begin{bmatrix} . & . & . & . & . & 1 \\ . & . & . & . & . & . \\ . & . & . & -1 & . & . \\ . & . & -1 & . & . & . \\ . & . & . & . & . & . \\ 1 & . & . & . & . & . \end{bmatrix},$$

$$A_3 = \begin{bmatrix} . & . & . & . & -1 & . \\ . & . & . & 1 & . & . \\ . & 1 & . & . & . & . \\ -1 & . & . & . & . & . \\ . & . & . & . & . & . \end{bmatrix},$$

are symmetric matrices, and so the first-order system is hyperbolic.

In order to describe difference methods for the numerical solution of (22), the region $-\infty < x, y < +\infty$, $t \geq 0$ is covered by a rectangular grid with lines parallel to the x-, y-, and t-axes. A typical grid point is given by $x = lh$, $y = mh$, $t = nk$ where h and k are grid spacings in the distance and time co-ordinates respectively, and l, m, and n are integers. Only *explicit* methods are considered for the solution of (22). In particular the Lax–Wendroff method is given by

$$\mathbf{U}_{l,m}^{n+1} = [I - \tfrac{1}{2}pAH_x - \tfrac{1}{2}pBH_y + \tfrac{1}{2}p^2 A^2 \delta_x^2 + \tfrac{1}{2}p^2 B^2 \delta_y^2 +$$

$$\tfrac{1}{8}p^2 (AB + BA)H_x H_y] \mathbf{U}_{l,m}^n, \qquad (23)$$

and the leap frog method by

$$\mathbf{U}_{l,m}^{n+1} = \mathbf{U}_{l,m}^{n-1} - p(AH_x + BH_y)\mathbf{U}_{l,m}^n. \qquad (24)$$

Exercise

3. Show by using Taylor expansions about the grid point (lh, mh, nk), that formula (23) has second order accuracy in space and time.

A stability analysis of a hyperbolic system in two space dimensions is extremely

G

difficult, even with A and B constant, unless A and B commute which is almost never the case in practice. However, it has been proved by Lax and Wendroff, without requiring commutation, that if A and B are constant, and if

$$|\lambda_m| = \max_{A,B} \left[|\lambda_A|, \ |\lambda_B| \right]$$

where

$$|A - \lambda_A I| = 0, \qquad |B - \lambda_B I| = 0,$$

then the Lax–Wendroff scheme (23) is stable if

$$p|\lambda_m| \leqslant \frac{1}{2\sqrt{2}} . \tag{25}$$

This limit on the stability of the Lax–Wendroff method in two space dimensions is much more severe than the C.F.L. limit

$$p|\lambda_m| \leqslant 1.$$

This situation contrasts with the one space dimensional case where the Lax–Wendroff formula attains the C.F.L. limit for stability. Strang (1963, 1964) and Gourlay and Morris (1968) have shown how to modify the Lax–Wendroff method in two space dimensions so that the stability condition of the revised method reaches the C.F.L. limit. The rotated Richymyer splitting of the Lax–Wendroff (see section 4.7) also reaches the C.F.L. limit of stability. A full discussion of the C.F.L. condition for hyperbolic systems in two and more space dimensions can be found in Wilson (1972).

Exercise

4. Following the analysis of (11b), derive the amplification matrices

$$[I - p^2 \ \{A^2(1\text{-}\cos \theta) + B^2(1\text{-}\cos \phi) + \tfrac{1}{2}(AB + BA)\sin \theta \sin \phi\}] \ -$$

$$ip(A \sin \theta + B \sin \phi)$$

and

$$[I - p^2(A \sin \beta h + B \sin \gamma h)^2]^{1/2} - ip(A \sin \beta h + B \sin \gamma h)$$

for the Lax–Wendroff and leap frog methods respectively, where $\theta = \beta h$, $\phi = \gamma h$, with β and γ arbitrary real numbers.

Example 9 Consider the hyperbolic system

$$\frac{\partial \mathbf{u}}{\partial t} = \begin{bmatrix} -2 & 1 \\ 1 & -2 \end{bmatrix} \frac{\partial \mathbf{u}}{\partial x} + \begin{bmatrix} -1 & 0 \\ 0 & -\alpha \end{bmatrix} \frac{\partial \mathbf{u}}{\partial y},$$

where $\alpha > 0$ is a parameter. The theoretical solution $\mathbf{u} = \begin{bmatrix} u_1 \\ u_2 \end{bmatrix}$ is given by

$$u_1 = \sin(x - t) + \sin(y - t)$$

and

$$u_2 = \sin(x - t) + \cos(y - \alpha t).$$

Solve the system numerically using the Lax–Wendroff method (23) for the cases $\alpha = 1, 2$ in the rectangular parallelepiped $0 \leqslant x, y \leqslant 1, 0 \leqslant t \leqslant T$.

Table 2

N	Lax–Wendroff
10	$- 0.0002708$
20	$- 0.0008011$
30	$- 0.0007832$
40	$- 0.0009333$
50	$- 0.0010576$
60	$- 0.0010846$
70	$- 0.0011059$
80	$- 0.0011764$
90	$- 0.0012479$
100	$- 0.0013057$
110	$- 0.0013390$
120	$- 0.0013358$
130	$- 0.0013168$
140	$- 0.0012769$
150	$- 0.0012362$

It should be noted that the matrices A and B in this system are negative definite and also that $AB = BA$ if $\alpha = 1$, whereas $AB \neq BA$ if $\alpha = 2$. The theoretical solution is used to obtain adequate initial and boundary data so that the difference method under consideration can be used to obtain a solution in the region described.

For $\alpha = 1$, $h = 0.1$, and $p = 0.1$, method (23) was allowed to run for 150 time steps and the maximum error in u_1 after a number of time steps (N), is shown in table 2.

Exercises

5. Using condition (25), show that the Lax–Wendroff method for solving the hyperbolic system with $\alpha = 1, 2$ in Example 9 is stable provided $p < 0.12$.

6. Repeat the calculation of Example 9 for 150 time steps with $\alpha = 2$, $h = 0.1$ and $p = 0.3$.

[Answer: the maximum errors in u_1 are given in table 3 after a number of time steps (N).]

Table 3

N	Lax–Wendroff
10	$- 0.0007461$
20	$- 0.0010186$
30	$- 0.0010647$
40	$- 0.0018811$
50	$- 0.0022701$
60	$- 0.0299503$
70	$+ 0.1063796$
	Unstable
	thereafter

So far the matrices A and B have been assumed to be constant. If the matrices depend on x, y and t, the Lax–Wendroff method given by equation (23) becomes

$$U_{l,m}^{n+1} = [I + \tfrac{1}{2}pA_{l,m}^{n+1/2}(\Delta_x + \nabla_x) + \tfrac{1}{2}pB_{l,m}^{n+1/2}(\Delta_y + \nabla_y) +$$

$$\tfrac{1}{4}p^2(A_{l,m}^{n+1/2}\Delta_x A_{l,m}^{n+1/2}\nabla_x + A_{l,m}^{n+1/2}\nabla_x A_{l,m}^{n+1/2}\Delta_x) +$$

$$\tfrac{1}{4}p^2(B_{l,m}^{n+1/2}\Delta_y B_{l,m}^{n+1/2}\nabla_y + B_{l,m}^{n+1/2}\nabla_y B_{l,m}^{n+1/2}\Delta_y) +$$

$$\tfrac{1}{8}p^2(A_{l,m}^{n+1/2}B_{l,m}^{n+1/2} + B_{l,m}^{n+1/2}A_{l,m}^{n+1/2})(\Delta_x + \nabla_x)(\Delta_y + \nabla_y)]U_{l,m}^n \qquad (23a)$$

4.7 Conservation laws in two space dimensions.

We now consider a system of conservation laws in two space variables given by

$$\frac{\partial \mathbf{u}}{\partial t} + \frac{\partial \mathbf{f}}{\partial x} + \frac{\partial \mathbf{g}}{\partial y} = 0. \qquad (26)$$

This can be written as the quasilinear system of equations

$$\frac{\partial \mathbf{u}}{\partial t} + A(\mathbf{u})\frac{\partial \mathbf{u}}{\partial x} + B(\mathbf{u})\frac{\partial \mathbf{u}}{\partial y} = 0 \qquad (27)$$

where $A(\mathbf{u})$ and $B(\mathbf{u})$ are the Jacobians of $\mathbf{f}$ and $\mathbf{g}$ with respect to $\mathbf{u}$ respectively. The numerical solution of (26) is of great importance since the equation models many practical problems, particularly in fluid dynamics (see Chapter 6). The interested reader should consult Morton (1977), Turkel (1974), Gottlieb and Turkel (1974), and Wilson (1972) for fuller details. In our limited discussion we shall content ourselves with listing some of the more popular types of difference schemes for solving (26).

Two-step Lax–Wendroff schemes

These are schemes which reduce to the Lax–Wendroff method (23) when A and B are constant matrices and intermediate quantities are eliminated. They include

(i) Richtmyer scheme.

$$\bar{U}_{l,m} = \tfrac{1}{2}(\mu_x + \mu_y)U_{l,m}^n - \tfrac{1}{2}p(\delta_x F_{l,m}^n + \delta_y G_{l,m}^n)$$

$$U_{l,m}^{n+1} = U_{l,m}^n - p(\delta_x \bar{F}_{l,m} + \delta_y \bar{G}_{l,m})$$

$$(28)$$

(ii) Rotated Richtmyer scheme.

$$\bar{U}_{l,m} = \mu_x \mu_y U_{l,m}^n - \tfrac{1}{2}p(\mu_y \delta_x F_{l,m}^n + \mu_x \delta_y G_{l,m}^n)$$

$$U_{l,m}^{n+1} = U_{l,m}^n - p(\mu_y \delta_x \bar{F}_{l,m} + \mu_x \delta_y \bar{G}_{l,m})$$

$$(29)$$

(iii) MacCormack scheme (1969, 1971).

$$\bar{U}_{l,m} = U_{l,m}^n - p[\Delta_x F_{l,m}^n + \Delta_y G_{l,m}^n]$$

$$U_{l,m}^{n+1} = \tfrac{1}{2}(U_{l,m}^n + \bar{U}_{l,m}) - \tfrac{1}{2}p[\nabla_x \bar{F}_{l,m} + \nabla_y \bar{G}_{l,m}].$$

$$(30)$$

Three other variations of the MacCormack scheme are possible. If we denote (30) by $(F_x, F_y : B_x, B_y)$ where F and B relate to forward and backward differences respectively, the other variations are $(F_x, B_y : B_x, F_y)$, $(B_x, F_y : F_x, B_y)$ and $(B_x, B_y : F_x, F_y)$.

The above schemes are generalizations of (18), (19) and (20) in two space dimensions.

Strang time splitting schemes

A different concept of splitting in time in order to solve (26) was introduced by Strang (1963, 1964, 1968) and the more practical details as far as finite difference methods are concerned, can be found in Gourlay (1977). The method amounts to considering the linearized version of (26) in the form (22) which we write as

$$\frac{\partial \mathbf{u}}{\partial t} + (L_x + L_y)\mathbf{u} = 0,$$

so that from Taylor's theorem

$$\mathbf{u}^{n+1} = \exp\left[-k(L_x + L_y)\right]\mathbf{u}^n. \tag{31}$$

(i) *Strang I.* Here we write

$$\exp\left[-k(L_x + L_y)\right] = \tfrac{1}{2}\left[\exp\left(-kL_x\right)\exp\left(-kL_y\right)\right.$$

$$\left. + \exp\left(-kL_y\right)\exp\left(-kL_x\right)\right], \tag{32}$$

which is correct to $0(k^2)$. Each exponential on the right-hand side of (32) is now replaced by its second-order expansion in k. The evaluation of the right-hand side of (32) requires four applications of the Lax–Wendroff method as defined by (11). In fact we use the Richtmyer two-stage version (18) of the Lax–Wendroff method so that (31), combined with (32), produces an eight-stage scheme for the solution of (26). We write it out as follows

$$\mathbf{V}_{(1)} = \mu_y\mathbf{U}^n - \tfrac{1}{2}p\delta_y\mathbf{G}^n \qquad\qquad \mathbf{W}_{(1)} = \mu_x\mathbf{U}^n - \tfrac{1}{2}p\delta_x\mathbf{F}^n$$

$$\mathbf{V}_{(2)} = \mathbf{U}^n - p\delta_y\mathbf{G}_{(1)} \qquad\qquad \mathbf{W}_{(2)} = \mathbf{U}^n - p\delta_x\mathbf{F}_{(1)}$$

$$\mathbf{V}_{(3)} = \mu_x\mathbf{V}_{(2)} - \tfrac{1}{2}p\delta_x\mathbf{F}_{(2)} \qquad\qquad \mathbf{W}_{(3)} = \mu_y\mathbf{W}_{(2)} - \tfrac{1}{2}p\delta_y\mathbf{G}_{(2)} \tag{33}$$

$$\mathbf{V}_{(4)} = \mathbf{V}_{(2)} - p\delta_x\mathbf{F}_{(3)} \qquad\qquad \mathbf{W}_{(4)} = \mathbf{W}_{(2)} - p\delta_y\mathbf{G}_{(3)}$$

$$\mathbf{U}_{l,m}^{n+1} = \tfrac{1}{2}\left(\mathbf{V}_{(4)} + \mathbf{W}_{(4)}\right).$$

This scheme which is discussed in greater detail in Gourlay and Morris (1968) has been shown by Strang to be stable subject only to the C.F.L. condition

$$p\,|\lambda_m| \leqslant 1, \text{ where } |\lambda_m| = \max_{A,B}\left[|\lambda_A|, |\lambda_B|\right].$$

(ii) *Strang II.* This time

$$\exp\left[-k(L_x + L_y)\right] = \exp\left(-\frac{k}{2}L_x\right)\exp\left(-kL_y\right)\exp\left(-\frac{k}{2}L_x\right) \tag{34}$$

which is again correct to $0(k^2)$. Substituting (34) into (31) and proceeding as in Strang I, we get an alternative to the multistage scheme (33).

Exercise

7. *Write out the multistep formulation of Strang II and compare your result with that given in Gourlay and Morris (1968).*

High order schemes.

So far all the schemes discussed have been explicit second-order in space and time.

We now consider schemes which are fourth-order in space and second-order in time. The best known is that due to Kreiss and Oliger (1972) which is based on the leap frog method and for equation (26) takes the form

$$\mathbf{U}^{n+1} = \mathbf{U}^{n-1} - p[10\delta_x\mu_x - 4\delta_x\mu_x{}^3]\mathbf{F}^n + [10\delta_y\mu_y - 4\delta_y\mu_y{}^3]\mathbf{G}^n. \qquad (35)$$

Exercise

8. Show that the scheme (35) can be written in the form

$$\mathbf{U}^{n+1} = \mathbf{U}^{n-1} + \tfrac{1}{6}\, p[(F_{l+2,m}^n - F_{l-2,m}^n) - 8\,(F_{l+1,m}^n - F_{l-1,m}^n)$$

$$+ (G_{l,m+2}^n - G_{l,m-2}^n) - 8\,(G_{l,m+1}^n - G_{l,m-1}^n)]$$

and verify that when $\mathbf{f} = A\mathbf{u}$ and $\mathbf{g} = B\mathbf{u}$ with A and B constant matrices, the method is fourth-order accurate in space and second-order in time.

4.8 Dissipation and dispersion

We shall discuss dissipation and dispersion in the context of finite difference replacements of the scalar hyperbolic equation

$$\frac{\partial u}{\partial t} + a\,\frac{\partial u}{\partial x} = 0 \quad (a > 0), \qquad (36)$$

on the half-space $\{-\infty < x < \infty,\ t > 0\}$. If the initial data is developed in a Fourier expansion

$$u(x, 0) = \sum_j A_j e^{i\beta_j x},$$

then, by virtue of the linearity of the problem, the properties of the solution may be assessed from the propagation of a single Fourier mode $e^{i\beta x}$ (By using Fourier transforms, β is allowed to range over all real numbers). When $u(x, 0) = e^{i\beta x}$, the solution of (36) is given by

$$u(x, t) = e^{i\beta(x - at)} \qquad (37)$$

and so each mode is transported with unit amplitude at a constant speed a independent of the frequency β. We now investigate the corresponding properties of finite difference replacements of (36). The initial step in the analysis follows the von Neumann method for stability where we assume that the finite difference approximation U_m^n may be expressed as

$$U_m^n = \xi^n\, e^{i\beta x m}. \qquad (38)$$

Substitution of this expression into the appropriate difference equation yields an expression for the amplification factor ξ. It is convenient to write $\xi = |\xi| e^{-i\beta\alpha k}$ so that (38) becomes

$$U_m^n = |\xi|^n\, e^{i\beta(x_m - \alpha t_n)}, \quad (t_n = nk), \tag{39}$$

which allows a more direct comparison to be made with the exact solution (37).

If the amplification factor ξ has modulus greater than unity the method is unstable whereas, if

$$|\xi| \leqslant 1 - \sigma(\beta h)^{2s} \qquad |\beta h| \leqslant \pi), \tag{40}$$

for some positive constant σ and positive integer s, the corresponding finite difference method is said to be *dissipative of order 2s*. Some measure of dissipation is often essential to stabilize an otherwise unstable method.

Example 10. Show that the Lax–Wendroff method (3) is dissipative of order four.

The modulus of the amplification factor ξ was earlier shown to be

$$|\xi| = [1 - 4r^2(1-r^2)\sin^4\frac{\beta h}{2}]^{1/2} \quad (r = pa).$$

For $r^2 \leqslant 1$ this expression decreases monotonically from the value $|\xi| = 1$ at $\beta h = 0$ to $|\xi| = |1 - 2r^2|$ at $\beta h = \pi$. For small values of βh, we find by Taylor series that

$$|\xi| \approx 1 - \tfrac{1}{2}r^2(1-r^2)(\beta h)^4$$

and the order of dissipation cannot therefore be less than four. The inequality (40) can now be satisfied by choosing, for example, $\sigma = \tfrac{1}{2}r^2(1-r^2)/\pi^4$.

Exercise

9. Show that the amplification factor for the leap frog method (4) satisfies $|\xi| = 1$ so that the method is not dissipative.

In view of (39) and (40), the solutions of dissipative schemes decay to zero as $n \to \infty$ and so they may not be appropriate for integrating hyperbolic equations over long time intervals. It is also clear from (40) that the decay of high frequency modes is much greater than that of low frequencies. This means that, for general initial data, the numerical solution will be smoothed as well as being damped as it evolves in time.

The dispersion properties of difference approximations relate to the speed of propagation of the numerical solution. Referring to the representation (39) of the numerical solution, the underlying method is said to be *dispersive* if the speed of propagation α depends on the frequency β. Again using the Lax–Wendroff method for illustration, the amplification factor ξ may be expressed in the form $|\xi| e^{-i\beta\alpha k}$ provided α is chosen to satisfy

$$\tan (\alpha\beta k) = \frac{r \sin \beta h}{1-2r^2 \sin^2 \tfrac{1}{2} \beta h}. \tag{41}$$

Clearly α depends on the frequency β and the method is therefore dispersive, in fact for $\beta h \approx \pi$ we see that $\alpha \simeq 0$ and consequently the high frequency waves are nearly stationary. For low frequency waves βh is small and Taylor series expansions can be used to estimate α. Using the approximations $\sin \beta h \approx \beta h - \tfrac{1}{6} (\beta h)^3$ and $(1 - x)^{-1} \approx 1 + x$, it follows from (41) that

$$\tan (\alpha\beta k) \approx r\beta h \ [1 + \tfrac{1}{6} (\beta h)^2 \ (3r^2 - 1)] \ .$$

Now since $\tan^{-1} x \approx x - \tfrac{1}{3}x^3$, we obtain

$$\alpha\beta k \approx r\beta h [1 - \tfrac{1}{6} (\beta h)^2 \ (1 - r^2)]$$

which leads to

$$\alpha \approx a [1 - \tfrac{1}{6} (\beta h)^2 (1 - r^2)] \ .$$

This shows that $\alpha < a$ for low frequency modes and the numerical solution therefore lags behind the true solution; that is, there is a *phase error*. The magnitude of the phase error increases with frequency and results in a loss of accuracy which is particularly noticeable when the initial data is in the form of a sharp signal (for instance a step function). Dispersive effects are usually less noticeable in dissipative schemes since the modes with greatest phase error are those which are mostly heavily damped. For this reason it is often beneficial to introduce dissipative mechanisms into dispersive difference schemes. For further information the reader should consult Richtmyer and Morton (1967), Morton (1977) and the references therein.

4.9 Stability of Initial Boundary Value Problems

So far in this chapter only the method of von Neumann, based on Fourier analysis, has been given as a means of analysing the stability of a hyperbolic system. This method gives necessary and sufficient conditions only for periodic or pure initial value problems, and so alternative methods are sought for mixed initial boundary value problems. These are usually energy methods (Richtmyer and Morton (1967) or methods based on spectral analysis. The latter, initially due to Godunov and Ryabenkii (1963), have been extended and improved by a number of people (see for instance Kreiss (1968), Gustafsson et al (1972), Morton (1977) and Burns (1978)).

An added complication is that the difference scheme being used often requires boundary conditions which are not given in the original problem. Unless great care is taken these extraneous boundary conditions can lead to instabilities in the difference calculation.

Example 11. Use the Lax–Wendroff method (11) to solve the system

$$\frac{\partial}{\partial t}\begin{pmatrix} u \\ v \end{pmatrix} + \begin{pmatrix} 0 & -1 \\ -1 & 0 \end{pmatrix} \frac{\partial}{\partial x}\begin{pmatrix} u \\ v \end{pmatrix} = 0 \qquad (x \geqslant 0, t > 0), \qquad (42)$$

subject to the conditions $u(0, t) = 0, u(x, 0) = f(x)$ and $v(x, 0) = g(x)$. (The properties of this system are described in Example 5).

Assuming for notational convenience that $\mathbf{U}_m^n$ has components U_m^n and V_m^n, equation (11) gives

$$U_m^{n+1} = U_m^n + \tfrac{1}{2}p\,(V_{m+1}^n - V_{m-1}^n) + \tfrac{1}{2}p^2(U_{m+1}^n - 2U_m^n + U_{m-1}^n) \qquad (43)$$

and

$$V_m^{n+1} = V_m^n - \tfrac{1}{2}p\,(U_{m+1}^n - U_{m-1}^n) + \tfrac{1}{2}p^2(V_{m+1}^n - 2V_m^n + V_{m-1}^n) \qquad (44)$$

for $n = 0, 1, \ldots$. Now suppose that the values U_m^n, V_m^n, $m = 0, 1, 2, \ldots$ are known at the time level $t = nk$ $(n \geqslant 0)$. The value $U_0^{n+1} = 0$ is deduced from the specified boundary condition at $x = 0$ and the quantities U_m^{n+1}, V_m^{n+1} $(m = 1, 2, 3, \ldots)$ computed from equations (43) and (44). The remaining unknown at the time level $t = (n + 1)k$ is therefore V_0^{n+1}. The additional equation necessary to determine V_0^{n+1} can be obtained from appropriate finite difference replacements of one or other of the component equations in (42). A number of different boundary approximations will now be described.

In view of the boundary condition $u(0, t) = 0(t > 0)$, we have $\partial u(0, t)/\partial t = 0$ and consequently, from the first of equations (42),

$$\frac{\partial v}{\partial x}(0, t) = 0 \qquad (t > 0). \qquad (45)$$

Replacing this condition by its forward space difference at $t = (n + 1)k$ leads to

$$V_1^{n+1} - V_0^{n+1} = 0 \qquad (46)$$

with a truncation error of $0(h)$. Alternatively, using Taylor expansions, it can be shown that an $0(h^2)$ one-sided difference replacement of this condition is given by

$$3V_0^{n+1} - 4V_1^{n+1} + V_2^{n+1} = 0. \qquad (47)$$

The second component of (42) reads

$$\frac{\partial v}{\partial t} - \frac{\partial u}{\partial x} = 0 \qquad (48)$$

and replacing the derivatives by their forward differences at $x = 0, t = nk$ leads to

$$V_0^{n+1} = V_0^n + p(U_1^n - U_0^n). \qquad (49)$$

Finally we derive a boundary approximation to (48) by the 'box integration' method. Integrating (48) over the cell PQRS shown in figure 2 gives

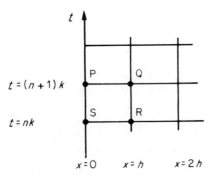

Figure 2

$$\int_{x=0}^{h} [v(x, (n + 1)k) - v(x, nk)]\,dx = \int_{t=nk}^{(n+1)k} [u(h, t) - u(0, t)]\,dt.$$

If the integrals are approximated by the trapezoidal rule we obtain

$$V_0^{n+1} = V_0^n + V_1^n - V_1^{n+1} + p[U_1^{n+1} + U_0^{n+1} - U_1^n - U_0^n]. \tag{50}$$

Further details of these and other boundary approximations may be found in Chu and Sereny (1974) and May (1978). The stability of (43) and (44), with additional boundary conditions given by one of (46), (47), (49) or (50), has been established, subject only to the C.F.L. condition $p \leqslant 1$, by Sundström (1975) and May (1978) using techniques based on the work of Godunov and Ryabenkii (1963).

Generally speaking, the errors due to approximate or additional boundary conditions will be represented by modes which are not of Fourier type. When the differential system is replaced by a dissipative difference scheme then the errors due to boundary conditions will either

(a) be confined to a few mesh lengths of the boundary if all the characteristics are outward-going, or
(b) be damped as they are propagated along inward-going characteristics.

The situation for non-dissipative difference schemes is much more delicate (see Kreiss and Lundqvist (1968), Gustafsson et al. (1972)).

4.10 Non-linear instability

To illustrate the effect that non-linear terms can have on the stability of hyperbolic finite difference equations we consider the solution of the equation

$$\frac{\partial u}{\partial t} + u\,\frac{\partial u}{\partial x} = 0 \qquad (x \geqslant 0, t > 0), \tag{51}$$

using the leap frog method (see equation (4))

$$U_m^{n+1} = U_m^{n-1} - pU_m^n (U_{m+1}^n - U_{m-1}^n).$$ (52)

We assume that initial data is available for $n = 0$ and $n = 1$ and that $u(0, t) = 0$ leading to $U_0^n = 0$ $(n \geqslant 0)$. In a local, linearized stability analysis, the non-linear coefficient U_m^n is 'frozen' at some value $\bar{U}$ and, since the leap frog method is stable for linear problems provided the C.F.L. condition is satisfied, we obtain $0 < p|\bar{U}| \leqslant 1$, that is

$$0 < pU_m^n \leqslant 1.$$ (53)

This is then used as a guide to the stability of the scheme (52). To demonstrate the inadequacy of this result it is sufficient to show that the method is unstable for one particular choice of initial data. Following Fornberg (1973), choose

$$U_m^0 = (-1)^{m+1} \epsilon \sin (\tfrac{1}{3}m\pi) \quad (m \geqslant 0),$$

and $U_m^1 = \alpha U_m^0$, where $\alpha > 1$ and ϵ (> 0) is sufficiently small so as not to violate the assumptions under which (52) was linearized. It can now be shown that there is a solution of (52) in the form

$$U_m^n = c_n (-1)^{m+1} \epsilon \sin \tfrac{1}{3}n\pi$$ (54)

provided c_n satisfies the recurrence relation

$$c_{n+1} - c_{n-1} = \gamma c_n^2 \quad (n \geqslant 1),$$

where $c_0 = 1$, $c_1 = \alpha$ and $\gamma = 2\epsilon p/\sqrt{3}$. Using an inductive argument, it can now be shown that

$$c_{2n} \geqslant (1 + \gamma)^n \text{ and } c_{2n+1} \geqslant (1 + \gamma)^n \quad (n \geqslant 0),$$

and so $c_n \to \infty$ as $n \to \infty$. We conclude from (54) that the method is therefore unstable for every choice of p and ϵ (no matter how small).

Since $u\partial u/\partial x = \tfrac{1}{2}\partial(u^2)/\partial x$, equation (51) may be written in the alternative form

$$\frac{\partial u}{\partial t} + \tfrac{1}{3}u\frac{\partial u}{\partial x} + \tfrac{1}{3}\frac{\partial u^2}{\partial x} = 0$$ (55)

for which the leap frog method becomes

$$U_m^{n+1} = U_m^{n-1} + \tfrac{1}{3}p\{ U_m^n(U_{m+1}^n - U_{m-1}^n) + (U_{m+1}^n)^2 - (U_{m-1}^n)^2 \}.$$ (56)

$$= U_m^{n-1} + \tfrac{1}{3}p(U_{m+1}^n + U_m^n + U_{m-1}^n)(U_{m+1}^n - U_{m-1}^n).$$

Equation (55) is known as the conservation form of (51) and its discrete form (56) is much less susceptible to the non-linear instability experienced by (52). This example demonstrates the care which has to be exercised in the approximation of non-linear terms in hyperbolic equations.

4.11 Second-order equations in one space dimension

A general second-order linear partial differential equation in two independent variables x, t can be written in the form

$$a\frac{\partial^2 u}{\partial t^2} + 2b\frac{\partial^2 u}{\partial t\partial x} + c\frac{\partial^2 u}{\partial x^2} + d\frac{\partial u}{\partial t} + e\frac{\partial u}{\partial x} + fu = g, \tag{57}$$

where the coefficients $a, b, c, \ldots, g$ are functions of the independent variables x and t. Equation (57) is said to be *hyperbolic* if

$$b^2 - ac > 0.$$

The simplest hyperbolic equation of this type is the wave equation

$$\frac{\partial^2 u}{\partial t^2} - \frac{\partial^2 u}{\partial x^2} = 0, \tag{58}$$

and this will now be studied in some detail to illustrate the type of initial and boundary conditions required to determine a unique solution of an hyperbolic equation. A solution of equation (58) can be written in the form

$$u = F(x + t) + G(x - t), \tag{59}$$

where F, G are arbitrary differential functions.

Now *Cauchy's initial value problem* consists of equation (58) together with the initial conditions $u(x, 0) = f(x)$ and $\partial u/\partial t(x, 0) = g(x)$, for $-\infty < x < +\infty$. It is easy to show that the solution of this problem is given by

$$u(x, t) = \frac{1}{2}[f(x + t) + f(x - t) + \int_{x-t}^{x+t} g(\xi)d\xi]. \tag{60}$$

This is one of the few non-trivial initial or boundary value problems which can be solved in a comparatively simple manner.

It is worth interpreting equation (60) in the light of the diagram in figure 3. Here the value of the solution at the point (x_0, t_0) depends on the initial data on the part of the x-axis between A and B, the points where the lines $x \pm t = $ constant through P cut the x-axis. The lines $x \pm t = $ constant are called the *characteristic curves* of the wave equation and the triangle PAB is called the *domain of dependence* of the point (x_0, t_0). These concepts generalize to all hyperbolic problems involving two independent variables, and the method of characteristics (see Garabedian (1964),

194

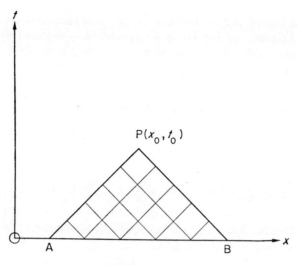

Figure 3

Jeffrey and Tanuiti (1964), Courant and Hilbert (1962), etc.) is undoubtedly the most effective method for solving hyperbolic problems in one space dimension. It might appear, therefore, that there is no role for finite difference methods to play in solving hyperbolic equations involving two independent variables. Nevertheless, it is necessary to consider finite difference methods for hyperbolic equations in one space dimension in order to instigate finite difference methods for hyperbolic equations in higher dimensions, where characteristic methods are less satisfactory.

Returning to the Cauchy problem involving the wave equation (58), the standard difference replacement of equation (58) on a rectangular grid is

$$\frac{1}{k^2}(U_m^{n+1} - 2U_m^n + U_m^{n-1}) - \frac{1}{h^2}(U_{m+1}^n - 2U_m^n + U_{m-1}^n) = 0, \qquad (61)$$

where k and h are the grid sizes in the time and distance co-ordinates respectively, and the difference formula (61) refers to the grid point $x = mh$, $t = nk$, where m and n are integers. The above formula can be rewritten in the explicit form

$$U_m^{n+1} = 2(1 - p^2)U_m^n + p^2(U_{m+1}^n + U_{m-1}^n) - U_m^{n-1},$$

where $p = k/h$ is the mesh ratio. If this formula is expanded by Taylor's theorem about the grid point (mh, nk) the result

$$k^2\left(\frac{\partial^2 u}{\partial t^2} - \frac{\partial^2 u}{\partial x^2}\right) + \frac{1}{12}k^2 h^2(p^2 - 1)\frac{\partial^4 u}{\partial x^4}$$

$$+ \frac{1}{360}k^2 h^4(p^4 - 1)\frac{\partial^6 u}{\partial x^6} + \ldots = 0$$

is obtained, and so the truncation error is

$$k^2 h^2 \left[\frac{1}{12} (p^2 - 1) \frac{\partial^4 u}{\partial x^4} + \frac{1}{360} h^2 (p^4 - 1) \frac{\partial^6 u}{\partial x^6} + \cdots \right].$$

The truncation error vanishes completely when $p = 1$, and so the difference formula

$$U_m^{n+1} = U_{m+1}^n + U_{m-1}^n - U_m^{n-1}$$

is an exact difference representation of the wave equation.

Assume that the functions f and g are prescribed on an interval $0 \leqslant x \leqslant 2Mh$ of the x-axis, where M is an integer. Then the initial conditions define U_m^0 and U_m^1 ($m = 0, 1, 2, \ldots, 2M$) in the following manner:

and
$$\left. \begin{aligned} U_m^0 &= f(mh) \\[4pt] U_m^1 &= (1 - p^2) f(mh) + \tfrac{1}{2} p^2 [f(m + 1)h + f(m - 1)h] + kg(mh) \\[4pt] & \quad (m = 0, 1, 2, \ldots, 2M). \end{aligned} \right\} \quad (62)$$

The latter formula is obtained from the Taylor expansion

$$u_m^1 = u_m^0 + k \left(\frac{\partial u}{\partial t} \right)_m^0 + \tfrac{1}{2} k^2 \left(\frac{\partial^2 u}{\partial t^2} \right)_m^0 + \cdots$$

$$= f(mh) + kg(mh) + \tfrac{1}{2} k^2 \left(\frac{\partial^2 u}{\partial x^2} \right)_m^0 + \cdots$$

$$\approx f(mh) + kg(mh) + \tfrac{1}{2} p^2 [f((m + 1)h) + f((m - 1)h) - 2f(mh)].$$

We now use formula (61) with $n = 1, 2, 3 \ldots$ to extend the initial values (62) to larger values of n. As illustrated in figure 4, the explicit calculation using formula (61) with $n = 1, 2, 3 \ldots$, together with values (62), comes to a halt at $n = N$ with the grid points, at which values of U have been calculated, lying inside a triangle. Hence the grid point Q, which is the apex of the triangle, has a *domain of dependence* on the x-axis which lies in the interval $0 \leqslant x \leqslant 2Mh$. This time the domain of dependence of the difference equation depends on the mesh ratio p and not on the characteristic curves through Q, as was the case with the differential equation. In this simple problem, it follows that three distinct cases arise for the difference problem:

(i) $0 < p < 1$. Here the domain of dependence for the difference equation includes that for the differential equation.

(ii) $p = 1$. The domains coincide, and the difference system takes the particularly simple form.

196

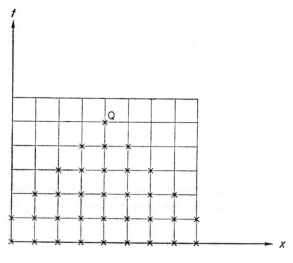

Figure 4

$$U_m^0 = f(mh)$$
$$U_m^1 = \tfrac{1}{2}[f((m+1)h) + f((m-1)h] + kg(mh) \left.\right\} \; (m = 0,1,\ldots,2M)$$
$$U_m^{n+1} = U_{m+1}^n + U_{m-1}^n - U_m^{n-1} \; (n = 1,2,\ldots,N).$$

(iii) $p > 1$. The domain of dependence for the difference equation lies inside that for the differential equation.

In each of the three cases, *provided p is kept constant*, the domain of dependence of the difference equation remains constant as h, $k \to 0$. The significance of the correspondence or otherwise of the domains of dependence of second-order hyperbolic equations and their associated explicit difference replacements is the substance of the Courant–Friedrichs–Lewy (C.F.L.) condition, which has already been discussed for first-order hyperbolic systems.

It is relatively easy to show that the domain of dependence for the difference equation must not lie inside that for the differential equation, i.e. $p \not> 1$. For if it does, it follows that the solution of the difference equation at a grid point $X = mh$, $T = nk$ is independent of the initial data of the problem which lies outside the domain of dependence of the difference equation but inside the domain of dependence of the differential equation. Alteration of this initial data will modify the solution of the differential equation, but leave the solution of the difference equation unaltered. This, of course, continues to apply when h, $k \to 0$ with p remaining constant and (X, T) remaining a *fixed* point. When $0 < p < 1$, the domain of dependence of the difference equation exceeds that of the differential equation, and it can be shown (Courant, Friedrichs and Lewy, (1928)), that the contribution to the solution of the difference equation at the grid point $X = mh$, $T = nk$ due to the extra initial data outside the domain of dependence of the differential equation tends to zero as h, $k \to 0$, p remaining constant and (X, T) remaining a *fixed* point. Thus the C.F.L. condition states that *the domain of dependence of the difference equation must include the domain of dependence of the differential equation.*

Example 12,. Solve the difference equation

$$U_m^{n+1} = U_{m+1}^n + U_{m-1}^n - U_m^{n-1}$$

on a square grid $(k = h)$ *in the* (x, t) *plane.*

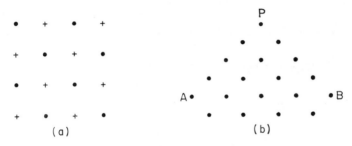

Figure 5

Consider the grid to be split up into two subgrids indicated by dots and crosses respectively as in figure 5(a). Since the difference equation connects the function values over each subgrid separately, we consider only the subgrid formed by the dots illustrated in figure 5(b). This is an arbitrary choice, and any result obtained can come equally from the subgrid formed by the crosses. Through a grid point P, the lines $x \pm t = $ constant meet the second row of dots at the points A and B respectively. The difference equation can be rewritten in the form

$$U_m^{n+1} - U_{m-1}^n = U_{m+1}^n - U_m^{n-1},$$

and if in figure 5 (b), we take $A \equiv (0, k)$, $B \equiv (8h, k)$ and $P \equiv (4h, 5k)$ where of course $k = h$, a simple calculation using the difference equation in this form gives

$$U_P - U_A = (U_P - U_3^4) + (U_3^4 - U_2^3) + (U_2^3 - U_1^2) + (U_1^2 - U_A)$$

$$= (U_B - U_7^0) + (U_6^1 - U_5^0) + (U_4^1 - U_3^0) + (U_2^1 - U_1^0),$$

where the right-hand side of this expression contains nodal values on the first and second rows only. Now let $h \to 0$, and the prescribed values on the first and second rows converge to a function $f(x)$, while the difference quotients of the type $(1/\sqrt{2}h)(U_B - U_7^0)$, $(1/\sqrt{2}h)(U_6^1 - U_5^0)$, etc., converge to the function $\gamma_+(x)$. If P remains fixed at the point (x, t) as $h \to 0$, then the result

$$u_P = f(x - t) + \frac{1}{\sqrt{2}} \int_{x-t}^{x+t} \gamma_+(\xi) d\xi$$

is obtained. Alternatively, by writing the difference equation in the form

198

$$U_m^{n+1} - U_{m+1}^n = U_{m-1}^n - U_m^{n-1},$$

we can obtain the result

$$u_P = f(x + t) + \frac{1}{\sqrt{2}} \int_{x-t}^{x+t} \gamma_-(\xi)d\xi.$$

where

$$\frac{1}{\sqrt{2}} [\gamma_+(\xi) + \gamma_-(\xi)] = g(\xi).$$

Hence by averaging the results obtained for u_P, we get

$$u_P = \frac{1}{2}[f(x + t) + f(x - t) + \int_{x-t}^{x+t} g(\xi)d\xi],$$

which corresponds exactly with equation (60) the solution of the differential equation.

It is instructive now to look at the stability of the difference equation (61). Following von Neumann we look at solutions of (61) which have the form

$$U_m^n = e^{\alpha nk} e^{i\beta mh} \tag{63}$$

where β is real and α is complex. Substituting (63) into (61) leads to

$$e^{2\alpha k} + 2(2p^2 \sin^2 \tfrac{1}{2}\beta h - 1)e^{\alpha k} + 1 = 0$$

which gives

$$e^{\alpha k} = 1 - 2p^2 \sin^2 \tfrac{1}{2}\beta h \pm 2p \sin \tfrac{1}{2}\beta h (p^2 \sin^2 \tfrac{1}{2}\beta h - 1)^{1/2}. \tag{64}$$

The von Neumann necessary condition for stability is

$$|e^{\alpha k}| \leqslant 1,$$

and so we examine the two cases

(i) $p^2 \sin^2 \tfrac{1}{2}\beta h \leqslant 1$. This gives

$$e^{\alpha k} = (1 - 2p^2 \sin^2 \tfrac{1}{2}\beta h) \pm i\, 2p \sin \tfrac{1}{2}\beta h (1 - p^2 \sin^2 \tfrac{1}{2}\beta h)^{1/2},$$

and so

$$|e^{\alpha k}| = 1.$$

(ii) $p^2 \sin^2 \frac{1}{2}\beta h > 1$. This leads to

$$|e^{\alpha k}| > 1,$$

from (64), if we consider the root with the negative sign.

Thus from (i), the difference equation (61) is stable for all β if $p \leqslant 1$.

Exercise

10. By writing (61) as the two-level system

$$U_m^{n+1} = 2(1-p^2)U_m^n + p^2(U_{m+1}^n + U_{m-1}^n) - V_m^n,$$

$$V_m^{n+1} = U_m^n,$$

obtain the amplification matrix, and hence show that (61) is stable if $p \leqslant 1$.

A more general hyperbolic equation in one space variable is

$$\frac{\partial^2 u}{\partial t^2} = \frac{\partial}{\partial x}\left(a(x,\,t)\,\frac{\partial u}{\partial x}\right) + b(x,\,t)\,\frac{\partial u}{\partial x} + c(x,\,t)u, \tag{65}$$

where $a(x,\,t) > 0$. As in the section on parabolic equations, the self-adjoint second-order term $\partial/\partial x\,[a(x,\,t)\partial u/\partial x]$ is replaced by

$$A_{m+1}^n U_{m+1}^n - (A_{m+1}^n + A_m^n)U_m^n + A_m^n U_{m-1}^n,$$

where

$$A_m^n = \frac{1}{h}\left[\int_{x_{m-1}}^{x_m} \frac{dx}{a(x,\,nk)}\right]^{-1}.$$

This leads to the explicit difference replacement of equation (65) given by

$$U_m^{n+1} = k^2\left[\left(A_{m+1} + \frac{b_m^n}{2h}\right)U_{m+1}^n + \left(A_m - \frac{b_m^n}{2h}\right)U_{m-1}^n\right] +$$

$$[2 - k^2(A_{m+1} + A_m - c_m^n)]\,U_m^n - U_m^{n-1} + 0(k^4 + k^2 h^2). \tag{66}$$

An *implicit* difference approximation to (58) is

$$[U_{m+1}^{n+1} - 2U_m^{n+1} + U_{m-1}^{n+1}] + 2[U_{m+1}^n - 2U_m^n + U_{m-1}^n] +$$

$$[U_{m+1}^{n-1} - 2U_m^{n-1} + U_{m-1}^{n-1}] = \frac{4}{p^2}[U_m^{n+1} - 2U_m^n + U_m^{n-1}],$$

which is stable for all $p > 0$. This is most conveniently written in the form

$$- (U_{m+1}^{n+1} + U_{m-1}^{n+1}) + 2\left(1 + \frac{2}{p^2}\right) U_m^{n+1} = + 2\left[(U_{m+1}^n + U_{m-1}^n) -\right.$$

$$\left. 2\left(1 - \frac{2}{p^2}\right) U_m^n\right] + \left[(U_{m+1}^{n-1} + U_{m-1}^{n-1}) - 2\left(1 + \frac{2}{p^2}\right) U_m^{n-1}\right], \qquad (67)$$

and a method based on formula (67) requires the solution of a tridiagonal system at each time step. This is easily accomplished using the technique outlined in section 2.5 of chapter 2.

The consistency of a finite difference approximation to a hyperbolic equation is now mentioned briefly. A difference approximation to a hyperbolic equation is *consistent* if

$$\frac{\text{Truncation error}}{k^2} \to 0 \text{ as } h, k \to 0.$$

Exercise

11. Show that formula (67) is a consistent approximation to the wave equation.

4.12 Second-order equations in two space dimensions

An arbitrary homogeneous *linear* hyperbolic equation of second order in *two space variables* can be transformed into the differential equation

$$\frac{\partial^2 u}{\partial t^2} - \left(a\frac{\partial^2 u}{\partial x_1^2} + 2b\frac{\partial^2 u}{\partial x_1 \partial x_2} + c\frac{\partial^2 u}{\partial x_2^2}\right)$$

$$+ d\frac{\partial u}{\partial t} + e\frac{\partial u}{\partial x_1} + f\frac{\partial u}{\partial x_2} + gu = 0, \qquad (68)$$

where the coefficients are functions of the independent variables x_1, x_2 and t and satisfy the conditions

$$a > 0, c > 0, \quad ac - b^2 > 0.$$

The most important example of equation (68) is the wave equation

$$\frac{\partial^2 u}{\partial t^2} - \frac{\partial^2 u}{\partial x_1^2} - \frac{\partial^2 u}{\partial x_2^2} = 0, \qquad (69)$$

and this will now be used to illustrate the connection between the domains of dependence of equation (69) and an explicit finite difference approximation to equation (69), with respect to the Cauchy initial value problem. It is conveneient to consider the latter to consist of equation (69) together with the initial conditions

$$u(x_1, x_2, 0) = f(x_1, x_2)$$

and
$$\left.\begin{array}{l}\\ \\ \frac{\partial u}{\partial t}(x_1, x_2, 0) = g(x_1, x_2)\end{array}\right\} \quad (-\infty < x_1, x_2 < +\infty).$$

The domain of dependence in the (x_1, x_2) plane of the differential equation (69) for a point $P(X_1, X_2, T)$ is the circle

$$(x_1 - X_1)^2 + (x_2 - X_2)^2 \leqslant T^2,$$

which is cut from the (x_1, x_2) plane by the circular cone with apex angle $\frac{1}{4}\pi$, vertex at P, and axis parallel to the t-axis. This cone is called the *characteristic cone* and plays a similar role to the characteristic lines $x \pm t = $ constant of the corresponding problem in one space dimension. The solution (60) for the latter problem also has a counterpart in two space dimensions, but the formula is now much more difficult (see Courant and Hilbert (1962), chapter 3, section 6.2).

The most natural difference approximation to equation (69) is

$$\frac{1}{k^2}(U_{l,m}^{n+1} - 2U_{l,m}^n + U_{l,m}^{n-1}) - \frac{1}{h^2}(U_{l+1,m}^n - 2U_{l,m}^n + U_{l-1,m}^n) -$$

$$\frac{1}{h^2}(U_{l,m+1}^n - 2U_{l,m}^n + U_{l,m-1}^n) = 0, \quad (70)$$

where k and h are the grid sizes in the time and distance co-ordinates respectively, and the difference formula (70) refers to the grid point

$$x_1 = lh, \quad x_2 = mh, \quad t = nk,$$

where l, m and n are integers. The above formula can be written in the explicit form

$$U_{l,m}^{n+1} = 2(1 - 2p^2)U_{l,m}^n +$$

$$p^2(U_{l+1,m}^n + U_{l-1,m}^n + U_{l,m+1}^n + U_{l,m-1}^n) - U_{l,m}^{n-1}, \quad (71)$$

where the mesh ratio $p = k/h$. If this formula is expanded by Taylor's theorem about the grid point (lh, mh, nk), it is easily shown that the principal part of the truncation error is given by

$$\frac{1}{12}k^2 h^2\left[(p^2 - 1)\left(\frac{\partial^4 u}{\partial x_1^4} + \frac{\partial^4 u}{\partial x_2^4}\right) + 2p^2\frac{\partial^4 u}{\partial x_1^2 \partial x_2^2}\right]. \quad (72)$$

Unlike the one space-dimensional case, there is no value of p which causes the truncation error to vanish identically and so there is no difference replacement of the form (71) which is an exact representation of the wave equation in two space dimensions.

The explicit difference formula (71) allows the value of the function u at a grid point $P(X_1, X_2, T)$ to be expressed uniquely in terms of the values of the function at certain points of two initial planes $t = 0, k$. These initial values are given by

$$U^0_{l,m} = f(lh, mh)$$

and

$$U^1_{l,m} = (1 - 2p^2) f(lh, mh) + \tfrac{1}{2}p^2 [f((l + 1)h, mh) + f((l - 1)h, mh) +$$

$$f(lh, (m + 1)h) + f(lh, (m - 1)h)] + kg(lh, mh),$$

respectively. The grid points which influence the value of U at the point P lie inside a pyramid which cuts out from the two initial planes $t = 0, k$ two rhombuses as domains of dependence. If $h, k \rightarrow 0$, with p remaining constant, the grid points which influence the value of U at $P(X_1, X_2, T)$, which remains fixed, continue to lie inside the above pyramid, and the domain of dependence of P is a rhombus cut out on the (x_1, x_2) plane by this pyramid. The Courant–Friedrichs–Lewy condition for the difference scheme (71) to be convergent for all smooth initial data is that the rhombus of dependence of the difference scheme must contain the circle of dependence of the differential equation in its interior. There is no loss of generality if we consider the grid point P to lie on the t-axis and so $X_1 = X_2 = 0$. If $T = nk$, then the domain of dependence for the differential equation is given by $x_1^2 + x_2^2 \leqslant n^2k^2$, and so in figure 6, $OQ = nk$. Now the rhombus of dependence for the difference equation, shown in figure 6, has $OR = OS = nh$, and so the C.F.L. condition is satisfied provided the rhombus includes the circle, and this is the case provided

$$nk \leqslant \frac{1}{\sqrt{2}} nh$$

or

$$p \leqslant \frac{1}{\sqrt{2}}. \tag{73}$$

This compares unfavourably with the C.F.L. condition of $p \leqslant 1$ for the one space-dimensional wave equation. In fact, it is easy to show that for the wave equation in s space dimensions, the C.F.L. condition for the explicit difference equation of the type (71) is

$$p \leqslant \frac{1}{\sqrt{s}},$$

a condition which grows more troublesome as the number of space dimensions increases.

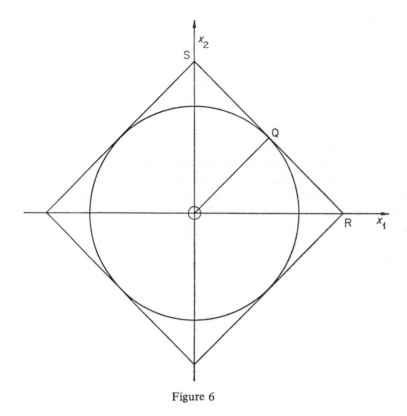

Figure 6

Condition (73) also guarantees the stability of formula (71). The proof of this is so similar to that given for the one space-dimensional case, that no details will be given.

Exercise

12. Using the standard nine point high accuracy difference replacement of

$$\frac{\partial^2 u}{\partial x_1^2} + \frac{\partial^2 u}{\partial x_2^2},$$

derive an explicit difference formula for the wave equation in two space dimensions. Determine the stability and C.F.L. conditions for this formula.

Once again it is possible to improve stability by considering implicit schemes. These implicit schemes will only be of value, however, if they can be used to calculate the solution in a relatively simple manner. In general, in two or more space dimensions this will not be the case, unless the difference operator at the most advanced time level can be factorized, and the difference equation rewritten as two or more simpler difference equations. This technique which was used with great success in parabolic equations can be employed with almost equal success in hyperbolic equations involving two or more space dimensions.

Consider an implicit difference formula in the form

$$[1 + a(\delta_{x1}^2 + \delta_{x2}^2) + d\delta_{x1}^2\delta_{x2}^2] U_{l,m}^{n+1} = [2 - b(\delta_{x_1}^2 + \delta_{x2}^2) -$$

$$e\delta_{x1}^2\delta_{x2}^2] U_{l,m}^{n} - [1 + c(\delta_{x1}^2 + \delta_{x2}^2) + f\delta_{x1}^2\delta_{x2}^2] U_{l,m}^{n-1}, \tag{74}$$

where the coefficients a, b, c, d, e, f are to be chosen so that formula (74) is an adequate replacement of equation (69), and can be used in a comparatively simple manner to calculate the solution. An implicit formula like (74), whether factorized or not, cannot be used to solve a pure initial value problem. It is, however, a natural formula to use for the solution of an initial boundary value problem where u and $\partial u/\partial t$ are given at $t=0$ for $0 \leqslant x_1$, $x_2 \leqslant 1$, and u is given on the four side boundaries $x_1 = 0, 1$ $(0 \leqslant x_2 \leqslant 1)$ and $x_2 = 0, 1$ $(0 \leqslant x_1 \leqslant 1)$ for $t > 0$. This is the problem of a vibrating square membrane fixed round its perimeter.

The difference operator at the advanced time level can be factorized if

$$d = a^2,$$

and formula (74) can be rewritten as

$$(1 + a\delta_{x_1}^2)(1 + a\delta_{x_2}^2)U^{n+1} = [2 - b(\delta_{x_1}^2 + \delta_{x_2}^2) - e\delta_{x_1}^2\delta_{x_2}^2]U^n -$$

$$[1 + c(\delta_{x_1}^2 + \delta_{x_2}^2) + f\delta_{x_1}^2\delta_{x_1}^2]U^{n-1}, \tag{75}$$

where U^s is written for $U_{l,m}^s$ $(s = n + 1, n, n - 1)$. The terms in formula (75) can be expanded in the Taylor series in terms of u and its derivatives at the grid point (lh, mh, nk), the even derivatives with respect to t being replaced by

$$\frac{\partial^2 u}{\partial t^2} = \frac{\partial^2 u}{\partial x_1^2} + \frac{\partial^2 u}{\partial x_2^2}, \quad \frac{\partial^4 u}{\partial t^4} = \frac{\partial^4 u}{\partial x_1^4} + 2\frac{\partial^4 u}{\partial x_1^2\partial x_2^2} + \frac{\partial^4 u}{\partial x_2^4}, \text{ etc.,}$$

from equation (69). The expansions up to and including terms involving h^6 are

$$u_{l,m}^{n+1} - 2u_{l,m}^n + u_{l,m}^{n-1} = p^2C_2 + \tfrac{1}{12}p^4C_4 + \tfrac{1}{6}p^4D_4 + \tfrac{1}{720}p^6C_6 + \tfrac{1}{240}p^6D_6,$$

$$(\delta_{x_1}^2 + \delta_{x_2}^2)u_{l,m}^{n\pm1} = C_2 \pm pC_3 + \tfrac{1}{2}(p^2 + \tfrac{1}{6})C_4 + p^2D_4 \pm$$

$$\tfrac{1}{6}p(p^2 + \tfrac{1}{2})C_5 \pm \tfrac{1}{3}p^3D_5 +$$

$$(\tfrac{1}{24}p^4 + \tfrac{1}{24}p^3 + \tfrac{1}{360})C_6 + \tfrac{1}{8}p^2(p^2 + \tfrac{1}{3})D_6,$$

$$\delta_{x_1}^2\delta_{x_2}^2 u_{l,m}^{n\pm1} = D_4 \pm pD_5 + \tfrac{1}{2}(p^2 + \tfrac{1}{180})D_6,$$

$$(\delta_{x_1}^2 + \delta_{x_2}^2)u_{l,m}^n = C_2 + \tfrac{1}{12}C_4 + \tfrac{1}{360}C_6,$$

$$\delta_{x_1}^2 \delta_{x_2}^2 u_{l,m}^n = D_4 + \tfrac{1}{360}D_6,$$

where

$$C_2 = h^2\left(\frac{\partial^2 u}{\partial x_1^2} + \frac{\partial^2 u}{\partial x_2^2}\right),$$

$$C_3 = h^3\frac{\partial}{\partial t}\left(\frac{\partial^2 u}{\partial x_1^2} + \frac{\partial^2 u}{\partial x_2^2}\right),$$

$$C_4 = h^4\left(\frac{\partial^4 u}{\partial x_1^4} + \frac{\partial^4 u}{\partial x_2^4}\right),$$

$$C_5 = h^5\frac{\partial}{\partial t}\left(\frac{\partial^4 u}{\partial x_1^4} + \frac{\partial^4 u}{\partial x_2^4}\right),$$

$$C_6 = h^6\left(\frac{\partial^6 u}{\partial x_1^6} + \frac{\partial^6 u}{\partial x_2^6}\right),$$

$$D_4 = h^4\frac{\partial^4 u}{\partial x_1^2 \partial x_2^2},$$

$$D_5 = h^5\frac{\partial^5 u}{\partial t \partial x_1^2 \partial x_2^2},$$

$$D_6 = h^6\frac{\partial^4}{\partial x_1^2 \partial x_2^2}\left(\frac{\partial^2 u}{\partial x_1^2} + \frac{\partial^2 u}{\partial x_2^2}\right).$$

Lees (1962) suggested formula (75) with

$$a = c = -\eta p^2, \quad b = -(1 - 2\eta)p^2, \quad e = -2\eta^2 p^4, \quad f = \eta^2 p^4,$$

where η is a parameter. Substitution of the above Taylor expansions leads to a principal truncation error of

$$-[\{(\eta - \tfrac{1}{12})p^4 + \tfrac{1}{12}p^2\} C_4 + 2(\eta - \tfrac{1}{12})p^4 D_4].$$

Fairweather and Mitchell (1965) suggested formula (75) with

$$a = c = \tfrac{1}{12}(1 - p^2), \quad b = -\tfrac{1}{6}(1 + 5p^2), \quad e = -\tfrac{1}{72}(1 + 10p^2 + p^4),$$

$$f = \tfrac{1}{144}(1 - p^2)^2.$$

This leads to a principal truncation error of

$$- \tfrac{1}{180} p^2 [(p^4 - \tfrac{3}{4})C_6 + \tfrac{7}{4}(p^4 - \tfrac{29}{21})D_6].$$

and so is an order h^2 more accurate than the method of Lees. A formula of intermediate accuracy was obtained by Samarskii (1964a), where the coefficients were

$$a = c = \tfrac{1}{12}(1 - 6p^2), \quad b = -\tfrac{1}{6}, \quad e = -\tfrac{1}{72}(1 + 36p^4),$$

$$f = \tfrac{1}{144}(1 - 6p^2)^2.$$

Exercise

13. Show that, in the methods of Lees, Samarskii, and Fairweather and Mitchell, the principal truncation errors are

$$0(h^4 + k^4), \quad 0(h^6 + k^4), \text{ and } 0(h^6 + k^6)$$

respectively.

The methods mentioned above have

$$b = -(2a + p^2), \quad c = a.$$

This guarantees an accuracy of at least $0(h^3 + k^3)$. Using these values, together with $f = a^2$, formula (75) can be rewritten in the form

$$(1 + a\delta_{x_1}^2)(1 + a\delta_{x_2}^2)[U^{n+1} + U^{n-1} - 2U^n] =$$

$$[p^2(\delta_{x_1}^2 + \delta_{x_2}^2) - (e + 2a^2)\delta_{x_1}^2 \delta_{x_2}^2]U^n, \tag{76}$$

and so the methods of Lees, Samarskii and Fairweather and Mitchell become

$$(1 - \eta p^2 \delta_{x_1}^2)(1 - \eta p^2 \delta_{x_2}^2)(U^{n+1} + U^{n-1} - 2U^n) = p^2(\delta_{x_1}^2 + \delta_{x_2}^2)U^n, \tag{76a}$$

$$[1 + \tfrac{1}{12}(1 - 6p^2)\delta_{x_1}^2][1 + \tfrac{1}{12}(1 - 6p^2)\delta_{x_2}^2](U^{n+1} + U^{n-1} - 2U^n) =$$

$$p^2[(\delta_{x_1}^2 + \delta_{x_2}^2) + \tfrac{1}{6}\delta_{x_1}^2 \delta_{x_2}^2]U^n \tag{76b}$$

and

$$[1 + \tfrac{1}{12}(1 - p^2)\delta_{x_1}^2][1 + \tfrac{1}{12}(1 - p^2)\delta_{x_2}^2](U^{n+1} + U^{n-1} - 2U^n) =$$

$$p^2[(\delta_{x_1}^2 + \delta_{x_2}^2) + \tfrac{1}{6}\delta_{x_1}^2 \delta_{x_2}^2]U^n \tag{76c}$$

respectively. The stability of these formulae is discussed in the respective articles.

We now rewrite (76) in the form

$$(1 + a\delta_{x_1}^2)(1 + a\delta_{x_2}^2)(U^{n+1} + U^{n-1} - 2U^n) = p^2 [\sum_i \delta_{xi}^2 + \beta\delta_{x_1}^2 \delta_{x_2}^2] U^n, \qquad (77)$$

where

$$\beta = -\frac{1}{p^2}(e + 2a^2).$$

Various splittings of formula (77) will be given which ease the computational procedure. The most obvious, but least economical, splitting from the point of view of computation is that of D'Yakonov and is given by

$$\left. \begin{array}{l} (1 + a\delta_{x_1}^2)U^{n+1\,*} = p^2 [\sum_i \delta_{xi}^2 + \beta\delta_{x_1}^2 \delta_{x_2}^2] U^n + \\ \qquad\qquad (1 + a\delta_{x_1}^2)(1 + a\delta_{x_2}^2)(2U^n - U^{n-1}) \end{array} \right\} \qquad (78)$$

and

$$(1 + a\delta_{x_2}^2)U^{n+1} = U^{n+1\,*}$$

where $U^{n+1\,*}$ is an intermediate value. This requires the computation of $\delta_{x_1}^2 U, \delta_{x_2}^2 U,$ and $\delta_{x_1}^2 \delta_{x_2}^2 U$ at each grid point of the net. A more economical splitting due to Lees is

$$\left. \begin{array}{l} (1 + a\delta_{x_1}^2)(U^{n+1\,*} - 2U^n + U^{n-1}) = p^2 \left(\sum_i \delta_{xi}^2 - \frac{\beta}{a} \delta_{x_2}^2 \right) U^n \\ (1 + a\delta_{x_2}^2)U^{n+1} = U^{n+1\,*} + \left(2a + p^2 \frac{\beta}{a} \right)\delta_{x_2}^2 U^n - a\delta_{x_2}^2 U^{n-1}, \end{array} \right\} \qquad (79a)$$

which may be written in the simplified form

$$\left. \begin{array}{l} (1 + a\delta_{x_1}^2) U^{n+1\,*} = p^2 \left(\sum_i \delta_{xi}^2 - \frac{\beta}{a} \delta_{x_2}^2 \right) U^n \\ (1 + a\delta_{x_2}^2)(U^{n+1} - 2U^n + U^{n-1}) = U^{n+1\,*} + p^2 \frac{\beta}{a} \delta_{x_2}^2 U^n, \end{array} \right\} \qquad (79b)$$

where $U^{n+1\,*}$ is a different intermediate value. Two other possible splittings are

$$\left. \begin{array}{l} (1 + a\delta_{x_1}^2)U^{n+1\,*} = p^2 [\sum_i \delta_{xi}^2 + \beta\delta_{x_1}^2 \delta_{x_2}^2] U^n \\ (1 + a\delta_{x_2}^2)(U^{n+1} - 2U^n + U^{n-1}) = U^{n+1\,*}, \end{array} \right\} \qquad (80)$$

and

208

and

$$(1 + a\delta^2_{x_1})U^{n+1\,*} = \frac{p^2}{a}[-1 + (a-\beta)\delta^2_{x_2}]U^n$$

and   (81)

$$(1 + a\delta^2_{x_2})(U^{n+1} - 2U^n + U^{n-1}) = U^{n+1\,*} + \frac{p^2}{a}(1 + \beta\delta^2_{x_2})U^n,$$

respectively. Further factorizations are given in Gourlay (1977) and Mitchell (1971).

Exercise

14. Show that the elimination of $U^{n+1\,*}$ in equations (79b), (80) and (81) respectively recovers formula (77).

Example 13. Consider the wave equation

$$\frac{\partial^2 u}{\partial t^2} = \frac{\partial^2 u}{\partial x_1^2} + \frac{\partial^2 u}{\partial x_2^2},$$

together with the initial conditions

$$u = \sin \pi x_1 \sin \pi x_2,$$

$$\frac{\partial u}{\partial t} = 0,$$

for $0 \leqslant x_1, x_2 \leqslant 1, t = 0$, and the boundary condition

$$u = 0,$$

on the boundary of the unit square for $t > 0$. The theoretical solution of this problem is

$$u = \sin \pi x_1 \sin \pi x_2 \cos 2\pi t,$$

within the region $0 \leqslant x_1, x_2 \leqslant 1, t > 0$. Use the split method of Lees given by the equations of (79a) (or 79b) with $a = -\eta p^2$ and $\beta = 0$ to obtain a numerical solution of the above problem.

Table 4

$t \diagdown \eta$	¼	½	1
0.3	− 0.009353	− 0.018065	− 0.034928
0.6	− 0.010148	− 0.020010	− 0.040224
0.9	+ 0.025128	+ 0.047913	+ 0.090206
1.2	+ 0.037914	+ 0.074418	+ 0.148135
1.5	− 0.019930	− 0.035643	− 0.057948
1.8	− 0.069506	− 0.134848	− 0.261908
2.1	− 0.011155	− 0.027693	− 0.076422
2.4	+ 0.087278	+ 0.165541	+ 0.306223
2.7	+ 0.062022	+ 0.128138	+ 0.276733
3.0	− 0.076253	− 0.137287	− 0.224734

We choose $h = \frac{1}{11}$ (i.e. 100 internal grid points at each time level) and $p = 0.66$. This leads to $k = 0.06$, and for $\eta = \frac{1}{4}, \frac{1}{2}, 1$, the errors using the method of Lees are shown in table 4 at every fifth time step at one of the four grid points nearest to the centre of the unit square. In all calculations, the values of u at internal grid points on the plane $t = k$ were taken from the theoretical solution. These values are necessary to start the calculations.

Example 14. Solve numerically the problem specified in Example 2, using the method of the equations of (79a) with $\beta = \frac{1}{6}$ and $a = \frac{1}{12}(1 - p^2)$. These are the values recommended by Fairweather and Mitchell.

Again we choose $h = \frac{1}{11}$, and calculations are carried out for a range of time increments k, viz. 0.03, 0.05, 0.06, 0.075. These correspond to values of p of 0.33, 0.55, 0.66 and 0.825 respectively. Since $p \leqslant \sqrt{3} - 1$, for stability it is expected that the last calculation will be unstable. The errors are shown in table 5, and the increased accuracy over the calculations carried out in Example 13 is demonstrated.

Table 5

t \ k	0.03	0.05	0.06	0.075
0.3	− 0.000015	− 0.000012	− 0.000008	+ 0.000001
0.6	− 0.000015	− 0.000012	− 0.000008	+ 0.000001
0.9	+ 0.000039	+ 0.000031	+ 0.000021	− 0.000003
1.2	+ 0.000056	+ 0.000045	+ 0.000031	− 0.000005
1.5	− 0.000032	− 0.000026	− 0.000018	+ 0.000002
1.8	− 0.000103	− 0.000082	− 0.000058	− 0.000004
2.1	− 0.000011	− 0.000009	− 0.000007	− 0.000186
2.4	+ 0.000132	+ 0.000106	+ 0.000075	− 0.002775
2.7	+ 0.000085	+ 0.000068	+ 0.000048	− 0.040980
3.0	− 0.000121	− 0.000097	− 0.000068	− 0.609823

Finally compact implicit schemes fourth-order accurate in space and time have been constructed for the wave equation

$$\frac{\partial^2 u}{\partial t^2} = a \frac{\partial^2 u}{\partial x_1^2} + c \frac{\partial^2 u}{\partial x_2^2}$$

by McKee (1973) (a and c functions of x_1 and x_2 only) and Ciment and Leventhal (1975) (a and c functions of x_1, x_2 and t). These are generalizations to variable coefficients of the scheme devised by Fairweather and Mitchell (1967).

Chapter 5

The Galerkin Method

5.1 Introduction

One of the most widely used techniques for discretizing elliptic equations and the spatial components of time dependent partial differential equations is the Galerkin method. Its success as a computational procedure is almost entirely due to the use of piecewise polynomial trial spaces constructed by the finite element method. (See Mitchell and Wait (1977), and Fairweather (1978)). Although the latter is usually considered to be an alternative to the finite difference method, in this chapter we look upon the Galerkin method with piecewise trial space as a means of constructing finite difference equations. Unlike the Taylor series method of construction the Galerkin method often directly incorporates useful information about the problem. For example, the Galerkin solution of Laplace's equation is the best solution in the energy norm.

In order to describe the Galerkin method we require some notation:

IR^N Euclidean space,

$\mathbf{x} = (x_1, x_2, \ldots, x_N)$,

R Bounded domain with boundary ∂R.

L_2 space

$$w(\mathbf{x}) \in L_2 (R) \text{ if } \left[\int_R w^2 \, d\mathbf{x} \right]^{1/2} < \infty$$

Inner products
$$(w_1, w_2) = \int_R w_1 w_2 \, d\mathbf{x}$$
$$\langle w_1, w_2 \rangle = \int_{\partial R} w_1 w_2 \, d\sigma$$
$\left. \right\} \; w_1, w_2 \in L_2 (R)$

H^m space $\qquad (m = 0, 1, 2, \ldots)$

$\alpha = (\alpha_1, \alpha_2, \ldots, \alpha_N) \qquad (\alpha_i \text{ non-negative integers})$

$$D^\alpha = \frac{\partial^{|\alpha|}}{\partial x_1^{\alpha_1} \ldots \partial x_N^{\alpha_N}} \qquad \left(|\alpha| = \sum_{i=1}^{N} \alpha_i \right)$$

$w(\mathbf{x}) \in H^m(R)$ if $D^\alpha w \in L_2(R) \qquad (|\alpha| \leqslant m)$

$\in H_g^m(R)$ if (i) $D^\alpha w \in L_2(R) \quad (|\alpha| \leqslant m)$

and $\qquad$ (ii) $w(\mathbf{x}) = g$ on ∂R.

Note the special case $H^0 \equiv L_2$.

5.2 Elliptic equations

Now consider the elliptic equation

$$Au = f \tag{1}$$

in a closed domain $R \subset IR^N$, where A is an mth order differential operator and f is a given function of $\mathbf{x}$. The solution of (1) is equivalent to determining u so that

$$(Au - f, v) = 0 \tag{2}$$

for all functions v lying in $L_2(R)$. This equation requires $D^m u \in L_2(R)$, and from a computational point of view, it would be convenient to reduce the disparity between the continuity requirements of u and v. This can usually be accomplished by integrating by parts in one dimension or by using Green's theorem in higher dimensions. For example, for the Laplacian operator in IR^N,

$$A = -\nabla^2, \qquad \nabla \equiv \left(\frac{\partial}{\partial x_1}, \ldots, \frac{\partial}{\partial x_N} \right) \tag{3}$$

and Green's theorem takes the form

$$(\nabla w, \nabla v) = -(\nabla^2 w, v) + \langle \frac{\partial w}{\partial n}, v \rangle \tag{4}$$

where $w \in H^2(R)$, $v \in H^1(R)$, and $\partial/\partial n$ denotes differentiation in the direction of the outward normal to ∂R. Combining (2), (3) and (4), we obtain

$$(\nabla u, \nabla v) - \langle \frac{\partial u}{\partial n}, v \rangle = (f, v)$$

in which we require only $u, v \in H^1(R)$. When the Dirichlet boundary condition $u = g$ is specified on ∂R, the *weak solution* $u \in H_g^1(R)$ of (1) is sought so that

$$(\nabla u, \nabla v) = (f, v) \qquad (\forall v \in H_0^1(R)). \tag{5}$$

Since any solution of (5) must satisfy $u = g$ on ∂R, the latter is often referred to as an *essential* boundary condition; the corresponding functions v must all vanish on ∂R. When the boundary condition is of the form $(\partial u / \partial n) + \alpha u = g$ on ∂R, the weak solution u is now sought in $H^1(R)$ so that

$$(\nabla u, \nabla v) - \langle g - \alpha u, v \rangle = (f, v) \qquad (\forall v \in H^1(R)). \tag{5a}$$

The bilinear form $a(u, v)$ associated with the Laplacian operator is defined as

$$a(u, v) = (\nabla u, \nabla v),$$

and if we assume that a bilinear form can be analogously defined for the general equation (1), the weak solution of (1) is determined from

$$a(u, v) = (f, v) \tag{6}$$

where along with continuity requirements, it is understood that this equation must hold for all test functions v which vanish at boundaries where u satisfies a Dirichlet condition.

The *Galerkin approximation* U to the weak solution u is written in the form

$$U(\mathbf{x}) = \sum_{i=0}^{M} U_i \phi_i(\mathbf{x}) \tag{7}$$

where the *trial* (or basis) *functions* $\phi_i(\mathbf{x})$ $(i = 0, 1, \ldots, M)$ form a basis for an $(M + 1)$ dimensional space which we denote by S^{M+1}. From (6) and (7), we define the Galerkin approximation to be that element $U \in S^{M+1}$ which satisfies

$$a(U, \phi_j) = (f, \phi_j) \qquad (\forall \phi_j \in S^{M+1}). \tag{8}$$

When Dirichlet boundary conditions are imposed on u, these must also be satisfied (at least approximately) by the Galerkin approximation U and in this case equation (8) need only be satisfied for those *test functions* ϕ_j which vanish on the boundary. In this formulation the test and trial functions are identical but we will show later that it is sometimes advantageous to choose different trial and test functions.

The operator A is said to be *self-adjoint* if $(Au, v) = (u, Av)$ and *positive definite* if $(Au, u) \geqslant \sigma(u, u) > 0$ for some positive non-zero constant σ. When both these conditions are satisfied, we can introduce the *energy space* H_A of the operator A in which the inner product and norm are defined as

$$(u, v)_A = (Au, v) = a(u, v)$$

and

$$\|u\|_A = (u, u)_A^{1/2}$$

respectively. The best approximation of the form (7) to the theoretical solution u with respect to the energy norm is such that

$$(U - u, \phi_j)_A = 0 \qquad (\forall \phi_j \in S^{M+1})$$

which may be rewritten as

$$a(U, \phi_j) = a(u, \phi_j).$$

Therefore, provided ϕ_j is an allowable test function in (6), this leads to

$$a(U, \phi_j) = (f, \phi_j)$$

which is identical to (8). Thus the Galerkin appoximation is the *best approximation to u in the energy space*.

A fuller explanation of this introductory material can be found in chapter 3 of Mitchell and Wait (1977).

5.3 Two-point boundary value problems

We now give some examples to show how the Galerkin method (8) is used to construct difference approximations to specific differential equations. Two-point boundary value problems will be discussed initially over the range $0 \leqslant x \leqslant 1$, which is divided up by a uniform grid of spacing h. The grid points are given by

$$x = jh \qquad (j = 0, 1, \ldots, M),$$

where $Mh = 1$. Our main concern is the discretization of the differential equation, although boundary conditions will be incorporated when appropriate.

Example 1. $\qquad A \equiv \dfrac{d^2}{dx^2}, \ f = 1.$

From (8), the Galerkin approximation U satisfies

$$(U', \phi_j') = (1, \phi_j) \tag{9}$$

for all test functions ϕ_j from the appropriate space which depends on the boundary conditions. A dash denotes differentiation with respect to x. The basis functions are chosen to be linear 'hat' functions.

H

214

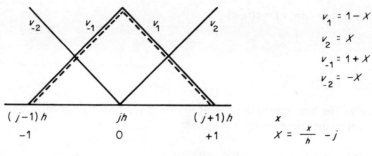

Figure 1

(a) *Linear 'hat' functions.* Here a typical test function (dotted line) along with the trial functions (full lines) required to evaluate (9) are shown in figure 1. The aproximation $U(x)$ is consequently a continuous piecewise linear function.

The internal grid points are given by $j = 1, 2, \ldots, M - 1$, and any *internal* grid point with its neighbours, one on either side, can be translated to the standard position shown in figure 1. In general, $X = (x/h) - j$, but when the inner product does not involve x explicitly, we can consider $j = 0$ for the general case, and so the evaluation of (9) proceeds as follows, where $U(X)$ and $\phi(X)$ are read from figure 1.

$-1 \leqslant X \leqslant 0$	$0 \leqslant X \leqslant +1$
$U(X) = -XU_{-1} + (1 + X)U_0$	$U(X) = (1 - X)U_0 + XU_{+1}$
$U'(x) = (-U_{-1} + U_0)\dfrac{1}{h}$	$U'(x) = (-U_0 + U_{+1})\dfrac{1}{h}$
$\phi(X) = 1 + X$	$\phi(X) = 1 - X$
$\phi'(x) = \dfrac{1}{h}$	$\phi'(x) = -\dfrac{1}{h}$

Since

$$(U', \phi_j') = \int_{x_{j-1}}^{x_{j+1}} U'(x)\, \phi_j'(x)\mathrm{d}x = \frac{1}{h}\int_{-1}^{+1} \frac{\mathrm{d}U(X)}{\mathrm{d}X}\,\frac{\mathrm{d}\phi_j(X)}{\mathrm{d}X}\,\mathrm{d}X,$$

equation (9) gives

$$\frac{1}{h}(-U_{j+1} + 2U_j - U_{j-1}) = h\left[\int_{-1}^{0}(1 + X)\,\mathrm{d}X + \int_{0}^{1}(1-X)\mathrm{d}X\right] = h, \quad (10)$$

at each *internal* node. This corresponds exactly to the standard finite difference replacement of $-u'' = 1$ at an internal node.

Example 2. Using the Galerkin method with linear 'hat' functions, obtain finite difference replacements of the differential systems

(i) $-u'' = x,$ $u'(0) = 0,\ u(1) = 1,$
(ii) $-u'' = x^2,$ $u = 0$ at $x = 0,1.$

In (i),

$$(-U'', \phi_j) = (x, \phi_j) \qquad (j = 0, 1, \dots, M).$$

Integration by parts (or using Green's theorem) leads to

$$-[U'\phi_j]_0^1 + (U',\phi_j') = (x, \phi_j) \qquad (j = 0, 1, \dots, M). \tag{11}$$

Now $U' = 0$ at $x = 0$, and $\phi_j = 0$ at $x = 1$ for $j = 0, 1, \dots, M - 1$.
Thus (11) reduces to

$$(U', \phi'_j) = (x, \phi_j) \qquad (j = 0, 1, \dots M - 1), \tag{12}$$

and

$$U = 1 \text{ at } x = 1.$$

Following the derivation of (10), we get from (12),

$$\frac{1}{h}(-U_{j+1} + 2U_j - U_{j-1}) = h^2 [\ \int_{j-1}^{j} (1 + X)(X + j)\, dX +$$

$$\int_{j}^{j+1} (1 - X)(X + j)\, dX\ ] = h^2 j \qquad (j = 1, 2, \dots, M - 1)$$

where again $X = (x/h) - j$, and

$$\frac{1}{h}(-U_{j+1} + U_j) = \tfrac{1}{6}h^2, \qquad (j = 0).$$

The discretized system is therefore

$$\frac{1}{h^2}
\begin{bmatrix}
1 & -1 & & & & \\
-1 & 2 & -1 & & & \\
& \cdot & \cdot & \cdot & & \\
& & \cdot & \cdot & \cdot & \\
& & & \cdot & \cdot & \cdot \\
& & & -1 & 2 & -1 \\
& & & & & 1
\end{bmatrix}
\begin{bmatrix}
U_0 \\ U_1 \\ U_2 \\ \cdot \\ \cdot \\ U_{M-1} \\ U_M
\end{bmatrix}
=
\begin{bmatrix}
\tfrac{1}{6}h \\ h \\ 2h \\ \cdot \\ \cdot \\ (M-1)h \\ 1/h^2
\end{bmatrix}
\tag{13}$$

(ii) is left as an exercise for the reader.

(b) *Quadratic functions.* Here typical test functions (dotted lines) at integer and half-integer nodes along with the trial functions (full lines) required to evaluate (9) are shown in figure 2. The approximation $U(x)$ is now a continuous piecewise quadratic function.

The internal integer grid points are given by $j = 1, 2, \ldots, M-1$ and any internal integer node can be translated into the standard position shown in figure 2. The evaluation of (9) at an *internal integer node* proceeds in the following manner, where $U(X)$ and $\phi(X)$ are read from figure 2.

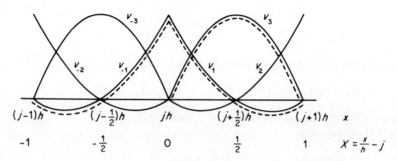

Figure 2

$$V_1 = 1 - 3X + 2X^2, \quad V_{-1} = 1 + 3X + 2X^2$$

$$V_2 = -X + 2X^2, \quad V_{-2} = X + 2X^2$$

$$V_3 = 4(X - X^2), \quad V_{-3} = -4(X + X^2)$$

$-1 \leqslant X \leqslant 0$	$0 \leqslant X \leqslant +1$
$U(X) = (X + 2X^2)U_{-1} - 4(X + X^2)U_{-1/2} + (1 + 3X + 2X^2)U_0$	$U(X) = (1 - 3X + 2X^2)U_0 + 4(X - X^2)U_{1/2} + (-X + 2X^2)U_1$
$U'(x) = [(1 + 4X)U_{-1} - 4(1 + 2X)U_{-1/2} + (3 + 4X)U_0]\dfrac{1}{h}$	$U'(x) = [(-3 + 4X)U_0 + 4(1 - 2X)U_{1/2} + (-1 + 4X)U_1]\dfrac{1}{h}$
$\phi(X) = 1 + 3X + 2X^2$	$\phi(X) = 1 - 3X + 2X^2$
$\phi'(x) = (3 + 4X)\dfrac{1}{h}$	$\phi'(x) = (-3 + 4X)\dfrac{1}{h}$

Hence (9) gives

$$\frac{1}{3h}(U_{j+1} - 8U_{j+1/2} + 14U_j - 8U_{j-1/2} + U_{j-1}) =$$

$$h\left[\int_{-1}^{0}(1 + 3X + 2X^2)\,dX + \int_{0}^{1}(1 - 3X + 2X^2)dX\right] = \frac{1}{3}h \tag{14}$$

at each *internal integer* node.

Any half integer node can be translated into the standard position $X = \frac{1}{2}$ as shown in figure 2. This time for $0 \leqslant X \leqslant 1$,

$$U(X) = (1 - 3X + 2X^2)U_0 + 4(X - X^2)U_{1/2} + (-X + 2X^2)U_1,$$

$$\phi(X) = 4(X - X^2),$$

and so (9) gives

$$\frac{8}{3h}(-U_{j+1} + 2U_{j+1/2} - U_j) = 4h \int_0^1 (X - X^2)\,dX = \frac{2h}{3} \qquad (15)$$

at a *half integer* node.

Exercise

1. Show that (14) reduces to (10) when (15) is used to eliminate $U_{j+1/2}$ and $U_{j-1/2}$ and comment on the relative accuracies of linear and quadratic basis functions when the Galerkin method is used to solve $-u'' = 1$.

The Galerkin approximation $U(x)$ can sometimes be expressed in terms of its derivatives as well as its function values at grid points. For instance, we can write

$$U(x) = \sum_{i=0}^{M} [U_i\,\phi_i\,(x) + U_i'\,\bar{\phi}_i\,(x)] \qquad (16)$$

where the basis functions $\phi_i(x)$ and $\bar{\phi}_i(x)$, $(i = 0, 1, \ldots, M)$ span a space $S^{2(M+1)}$. From (6) and (16) we define the Galerkin approximation to be the element $U \in S^{2(M+1)}$ which satisfies

$$\left.\begin{array}{l} a(U, \phi_j) = (f, \phi_j), \\[2mm] a(U, \bar{\phi}_j) = (f, \bar{\phi}_j). \end{array}\right\} \quad (\forall\,\phi_j,\,\bar{\phi}_j \in S^{2(M+1)}) \qquad (17)$$

where $S^{2(M+1)}$ must be compatible with the boundary conditions. We now return to Example 2 where $A \equiv -\,d^2/dx^2$, $f = 1$ and consider the basis functions to be *Hermite cubic functions*. These are illustrated in figure 3 for the standard position. In $-1 \leqslant X \leqslant 0$, the approximation is given by

$$U(X) = X^2(3 + 2X)U_{-1} + X^2(1 + X)\,U_{-1}' + (1 + X)^2\,(1 - 2X)U_0 + X(1 + X)^2\,U_0'$$

and the test functions by

$$\phi(X) = (1 + X)^2\,(1 - 2X)$$

$$\bar{\phi}(X) = X(1 + X)^2.$$

In $0 \leqslant X \leqslant +1$, we get

$$U(X) = (1 - X)^2 (1 + 2X)U_0 + X(1 - X)^2 U_0' + X^2 (3 - 2X)U_1 + X^2(X - 1)U_1'$$

and

$$\phi(X) = (1 - X)^2 (1 + 2X),$$

$$\bar{\phi}(X) = X(1 - X)^2,$$

respectively. Following evaluation of the inner products, (17) gives the two equations

$$\frac{6}{5h}(-U_{j+1} + 2U_j - U_{j-1}) + \tfrac{1}{10}(U_{j+1}' - U_{j-1}') =$$

$$h[\int_{-1}^{0}(1 + X)^2 (1 - 2X)dX + \int_{0}^{1}(1 - X)^2 (1 + 2X)dX] = h, \qquad (18)$$

and

$$-\frac{1}{10h}(U_{j+1} - U_{j-1}) + \tfrac{1}{30}(-U_{j+1}' + 8U_j' - U_{j-1}') =$$

$$h[\int_{-1}^{0} X(1 + X)^2 dX + \int_{0}^{1} X(1 - X)^2 dX] = 0,$$

which hold at each internal node.

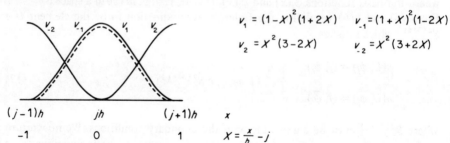

$$v_1 = (1-X)^2(1+2X) \qquad v_{-1} = (1+X)^2(1-2X)$$
$$v_2 = X^2(3-2X) \qquad v_{-2} = X^2(3+2X)$$

$$(j-1)h \qquad jh \qquad (j+1)h \qquad x$$
$$-1 \qquad 0 \qquad 1 \qquad X = \frac{x}{h} - j$$

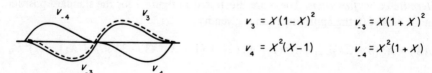

$$v_3 = X(1-X)^2 \qquad v_{-3} = X(1+X)^2$$
$$v_4 = X^2(X-1) \qquad v_{-4} = X^2(1+X)$$

Figure 3

Exercise

2. Use the Galerkin method with Hermite cubics and $h = \tfrac{1}{10}$ to solve the differential system

$$-u'' = x, \qquad u' = 0 \text{ at } x = 0, \quad u = 1 \text{ at } x = 1.$$

Compare the accuracy of the function and its derivative with the theoretical solution at grid points and elsewhere in the range. As an intermediate step, show that (18) becomes

$$\frac{6}{5h}(-U_1 + U_0) + \tfrac{1}{10}U_1' = \tfrac{3}{20}h$$

at $x = 0$, and

$$\frac{1}{10h}(-1 + U_{M-1}) + \tfrac{1}{30}(4U_M' - U_{M-1}') = (\tfrac{1}{30} - \frac{M}{12})h$$

at $x = 1$. [Reminder: Inner products involving x explicitly require the transformation $X = x/h - j$ where $j = 0, 1, \ldots, M$].

In the remainder of this section we investigate the finite difference replacements of first derivative terms generated by the Galerkin method using piecewise linear, quadratic and cubic basis functions.

Example 3. $A \equiv \dfrac{d}{dx}$, $f = 1$.

From (8), the Galerkin approximation U satisfies

$$(U', \phi_j) = (1, \phi_j) \qquad (\forall\ \phi_j \in S^{M+1}) \tag{19}$$

With linear 'hat' functions, (19) gives

$$\tfrac{1}{2}(U_1 - U_{-1}) = h,$$

and with quadratic functions,

$$-\tfrac{1}{6}U_1 + \tfrac{2}{3}U_{1/2} - \tfrac{2}{3}U_{-1/2} + \tfrac{1}{6}U_{-1} = \tfrac{1}{3}h$$

at an integer node, and

$$\tfrac{2}{3}(U_1 - U_0) = \tfrac{2}{3}h$$

at a half integer node, where the notation is for the standard placements of the nodes. If we use Hermite cubic functions, (19) gives

$$\tfrac{1}{2}(U_1 - U_{-1}) - \frac{h}{10}(U_1' - 2U_0' + U_{-1}') = h$$

and

$$\tfrac{1}{10}(U_1 - 2U_0 + U_{-1}) - \frac{h}{60}(U_1' - U_{-1}') = 0$$

respectively.

Example 4. $A \equiv -\dfrac{d^2}{dx^2} + K\dfrac{d}{dx}$, $f = 0$.

K is a positive constant and the boundary conditions are $u = 1$ at $x = 0$ and $u = 0$ at $x = 1$. The theoretical solution is $u = (e^{Kx} - e^K)/(1 - e^K)$. This example uses the results already obtained in Examples 2 and 3.

With linear basis functions, the Galerkin method gives

$$(1 - L)U_{j+1} - 2U_j + (1 + L)U_{j-1} = 0 \quad (j = 1, 2, \ldots, (M - 1)) \quad (20)$$

where $L = \frac{1}{2}hK$. With quadratic functions we get

$$(1 - L)U_{j+1} - 4(2 - L)U_{j+\frac{1}{2}} + 14U_j - 4(2 + L)U_{j-\frac{1}{2}} + (1 + L)U_{j-1} = 0$$

$$(j = 1, 2, \ldots, (M - 1)) \quad (21)$$

at the integer nodes, and

$$(1 - \tfrac{1}{2}L)U_j - 2U_{j-\frac{1}{2}} + (1 + \tfrac{1}{2}L)U_{j-1} = 0 \quad (j = 1, 2, \ldots, M) \quad (22)$$

at the half integer nodes. In both cases, of course, $U_0 = 1$ and $U_M = 0$.

Exercise

3. Obtain the theoretical solution of the difference equation (20) and show that it is oscillatory if $L > 1$. Also use (22) to eliminate $U_{j+\frac{1}{2}}$ and $U_{j-\frac{1}{2}}$ from (21), and show that the resulting three point difference formula has a solution which is oscillation free for all L.

[Hint: Assuming that (20) has a solution of the form $U_j = A\xi^j$, then ξ must satisfy the quadratic equation $(1 - L)\xi^2 - 2\xi + (1 + L) = 0$. If the roots of this equation are ξ_1 and ξ_2, the general solution is $U_j = A\xi_1^j + B\xi_2^j$ where the constants A and B can be determined by applying the boundary conditions.]

With Hermite cubic basis functions in example 4, the Galerkin method gives

$$(12 - 10L)U_{j+1} + (2L - 1)hU'_{j+1} - 24U_j - 4LhU'_j + (12 + 10L)U_{j-1}$$

$$+ (1 + 2L)hU'_{j-1} = 0$$

and $\quad (23)$

$$3(1 - 2L)U_{j+1} + (1 + L)hU'_{j+1} + 12LU_j - 8hU'_j - 3(1 + 2L)U_{j-1}$$

$$+ (1 - L)hU'_{j-1} = 0,$$

at the internal nodes $j = 1, 2, \ldots, (M - 1)$,

$$3(1 - 2L)U_1 + (1 + L)hU'_1 - 4hU'_0 = 3(1 - 2L)$$

at $x = 0$, and

$$-3(1 + 2L)U_{M-1} + (1 - L)hU'_{M-1} - 4hU'_M = 0$$

at $x = 1$.

5.4 The Galerkin method with different test and trial functions

In example 3, it was seen that the Galerkin finite element method with linear basis functions caused a first derivative to be replaced by the standard central difference formula. It may be asked if it is possible to obtain the standard backward (upwinded) and forward (downwinded) difference replacements of the first derivative by Galerkin methods. To accomplish this we use a generalization of the Galerkin method called the *Petrov–Galerkin* method (Anderssen and Mitchell (1979)), which for Example 3 is given by (19) modified to read: the Petrov–Galerkin approximation is the element $U \in \Phi^{M+1}$ which satisfies

$$(U', \psi_j) = (1, \psi_j) \quad (\forall \, \psi_j \in \Psi^{M+1}) \tag{24}$$

where the trial and test spaces Φ^{M+1} and Ψ^{M+1} respectively are different finite dimensional subspaces of $H^1(R)$. We choose the trial functions $\phi_i(x)$ ($i = 0, 1, \ldots, M$) to be the linear 'hat' functions described in (a) and the test functions to be

$$\psi_j(x) = \phi_j(x) + \alpha\sigma_2 \left(\frac{x}{h} - j \right) \quad (j = 0, 1, \ldots, M), \tag{25}$$

where

$$\sigma_2(X) = \begin{cases} 0 & |X| > 1 \\ -3X(1 - X) & 0 \leqslant X \leqslant 1 \\ -\sigma_2(-X) & -1 \leqslant X \leqslant 0 \end{cases} \tag{26}$$

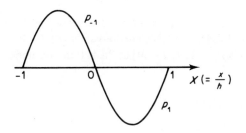

$$P_1 = -3X(1-X)$$

$$P_{-1} = -3X(1+X)$$

Figure 4

is a *quadratic* perturbing function shown in figure 4, and α is an arbitrary parameter. The evaluation of (24) now proceeds as follows:

$-1 \leqslant X \leqslant 0$	$0 \leqslant X \leqslant +1$
$U'(x) = (-U_{-1} + U_0)\dfrac{1}{h}$	$U'(x) = (-U_0 + U_1)\dfrac{1}{h}$
$\psi(X) = (1 + X) - 3\alpha X(1 + X)$	$\psi(X) = (1 - X) - 3\alpha X(1 - X)$

and so (24) gives

$$\tfrac{1}{2}[(1 - \alpha)U_1 + 2\alpha U_0 - (1 + \alpha)U_{-1}] = h. \tag{27}$$

This gives the forward, central, and backward difference replacements for the first derivative when $\alpha = -1, 0,$ and $+1$ respectively.

Exercise

4. Show that with the modified test functions given by (25) there is no change in the difference approximation to the equation $-u'' = 1$ in Example 2. Hence obtain the difference approximation

$$[(1 - L) + \alpha L]U_1 - 2(1 + \alpha L)U_0 + [(1 + L) + \alpha L]U_{-1} = 0 \tag{28}$$

at an internal node for the equation $-u'' + Ku' = 0$ of Example 4, where $L = \tfrac{1}{2}Kh$. Show also that when

$$\alpha = \coth L - \frac{1}{L}, \tag{29}$$

the theoretical solution of the difference equation (28) coincides with the exact solution of the differential equation of Example 4 with the given boundary conditions. This result was originally given by Il'in (1969), and the upwinding parameter α given by (29) has been used in solving convection diffusion problems by Heinrich et al. (1977).

The quadratic basis functions described in section 5.3 can also be upwinded using the Petrov–Galerkin method (24). The trial functions are $\phi_i(x)$ $(i = 0, 1, \ldots, M)$ at the integer nodes and $\phi_{i-\frac{1}{2}}(x)$ $(i = 1, 2, \ldots, M)$ at the half integer nodes. The test functions are chosen to be

$$\psi_j(x) = \phi_j(x) + \alpha_1 \sigma_3\left(\frac{x}{h} - j\right) + \alpha_2 \sigma_3\left(\frac{x}{h} - j - 1\right) \quad (j = 0, 1, \ldots, M) \tag{30}$$

at the integer nodes, and

$$\psi_{j-\frac{1}{2}}(x) = \phi_{j-\frac{1}{2}}(x) + 4\alpha_3 \sigma_3\left(\frac{x}{h} - j\right) \quad (j = 1, 2, \ldots, M) \tag{31}$$

at the half integer nodes, where

$$\sigma_3(X) = \begin{cases} -5X(X + \tfrac{1}{2})(X + 1) & -1 \leqslant X \leqslant 0 \\ 0 & \text{elsewhere} \end{cases}$$

is a *cubic* perturbing function, shown in figure 5, and α_i $(i = 1, 2, 3)$ are arbitrary parameters. For Example 3, the Petrov–Galerkin method gives

$$-\tfrac{1}{6}(1 - \alpha_2)U_1 + \tfrac{4}{6}(1 - \tfrac{1}{2}\alpha_2)U_{\frac{1}{2}} + \tfrac{1}{6}(\alpha_2 + \alpha_1)U_0 - \tfrac{4}{6}(1 + \tfrac{1}{2}\alpha_1)U_{-\frac{1}{2}} +$$

$$\tfrac{1}{6}(1 + \alpha_1)U_{-1} = \tfrac{1}{3}h$$

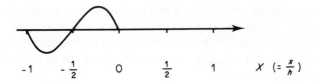

Figure 5

at an integer node, and

$$\tfrac{2}{3}(1 + \alpha_3)U_1 - \tfrac{4}{3}\alpha_3 U_{\frac{1}{2}} - \tfrac{2}{3}(1 - \alpha_3)U_0 = \tfrac{2}{3}h$$

at a half integer node. For Example 1, we still get the results given by (14) and (15).

Exercise

5. Show that in Example 4, the Petrov–Galerkin method using the test functions given by (30) and (31) leads to

$$(1 - L + \alpha_2 L)U_{j+1} - 4(2 - L + \tfrac{1}{2}\alpha_2 L)U_{j+\frac{1}{2}} + (14 + \alpha_2 L + \alpha_1 L)U_j$$
$$- 4(2 + L + \tfrac{1}{2}\alpha_1 L)U_{j-\frac{1}{2}} + (1 + L + \alpha_1 L)U_{j-1} = 0 \quad (32)$$

at the integer nodes $j = 1, 2, \ldots, M-1$, and

$$(1 - \tfrac{1}{2}L - \tfrac{1}{2}\alpha_3 L)U_j - (2 - \alpha_3 L)U_{j-\frac{1}{2}} + (1 + \tfrac{1}{2}L - \tfrac{1}{2}\alpha_3 L)U_{j-1} = 0 \quad (33)$$

at the half integer nodes $j = 1, 2, \ldots, M$. (Further details concerning upwinded quadratics can be found in Christie and Mitchell (1978).)

5.5 Two space variables

The Galerkin method (8) is now applied to the solution of elliptic partial differential equations in two space variables. In particular we look at the Poisson equation

$$-\left(\frac{\partial^2 u}{\partial x^2} + \frac{\partial^2 u}{\partial y^2} \right) = 1,$$

for which (8) becomes

$$\iint_R \left(\frac{\partial U}{\partial x} \frac{\partial \phi_j}{\partial x} + \frac{\partial U}{\partial y} \frac{\partial \phi_j}{\partial y} \right) dx\,dy = \iint_R \phi_j \, dx\,dy \quad (34)$$

where U is given by (7), and $\phi_j(x, y)$ $(j = 0, 1, \ldots, M)$ is a suitable set of basis functions. Equation (34) does not take account of boundary conditions and so the functions $\phi_j(x, y)$ $(j = 0, 1, \ldots, M)$ must all vanish on the boundary ∂R.

We assume first that the region R, over which the Poisson equation is satisfied, is covered by a *rectangular* grid of size h and that inside each rectangular element, a *bilinear* function applies which picks up the unknown function values at the corners of the rectangle. Away from the boundary, any *internal* grid point (lh, mh), $(l,m$ integers) and the four rectangles containing the point can be translated to the standard position shown in figure 6.

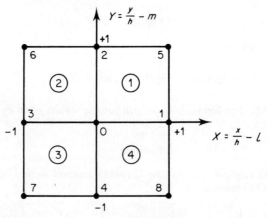

Figure 6

We number the elements 1 , 2 , 3 , and 4 , and the nodes 0, 1, 2, ..., 8. In region 1 , the approximant is

$$U(X, Y) = X(1 - Y)U_1 + XYU_5 + (1 - X)YU_2 + (1 - X)(1 - Y)U_0$$

$$\frac{\partial U}{\partial X} = (1 - Y)U_1 + YU_5 - YU_2 - (1 - Y)U_0$$

$$\frac{\partial U}{\partial Y} = -XU_1 + XU_5 + (1 - X)U_2 - (1 - X)U_0,$$

and the test function is

$$\phi(X, Y) = (1 - X)(1 - Y)$$

$$\frac{\partial \phi}{\partial X} = -(1 - Y), \quad \frac{\partial \phi}{\partial Y} = -(1 - X).$$

We now find, for region 1 , that

$$\iint_1 \left(\frac{\partial U}{\partial x} \frac{\partial \phi_j}{\partial x} + \frac{\partial U}{\partial y} \frac{\partial \phi_j}{\partial y} \right) dx \, dy = \frac{1}{6}(4U_0 - U_1 - U_2 - 2U_5)$$

and

$$\iint_1 \phi_j \, dx \, dy = \tfrac{1}{4}h^2.$$

Adopting a similar procedure for regions 2, 3 and 4 and summing the results leads to

$$\tfrac{1}{3}[8U_0 - \sum_{i=1}^{8} U_i] = h^2 \tag{35}$$

for the difference replacement of the Poisson equation at an internal grid point.

Exercise

6. Use the Galerkin method with bilinear basis functions on a rectangular grid to discretize the elliptic equation

$$-\left(\frac{\partial^2 u}{\partial x^2} + \frac{\partial^2 u}{\partial y^2} + \frac{\partial u}{\partial x}\right) = 1$$

at an internal grid point. [In the notation of figure 6 the result is

$$\frac{1}{3h^2}[8U_0 - \sum_{i=1}^{8} U_i] + \frac{1}{12h}[(U_5 + U_8 - U_6 - U_7) + 4(U_1 - U_3)] = 1]$$

We now subdivide the rectangular grid by a set of inclined parallel lines passing through the nodes. This results in a *triangular* grid and inside each triangular element a *linear* function applies which picks up the unknown function values at the vertices of each triangle. Away from the boundary, any internal grid point and the six triangles containing the point can be translated into the standard position shown in figure 7. We number the elements 1, 2, ..., 6 and the nodes 0, 1, 2, ..., 6, as shown in the figure.

The definition of the approximant $U(x, y)$ and the test function $\phi(x, y)$ in each of the elements 1, 2, ..., 6 is listed below.

Element number	Approximant $U(x, y)$	Test function $\phi(x, y)$
1	$(1 - X)U_0 + (X - Y)U_1 + YU_5$	$1 - X$
2	$(1 - Y)U_0 + (Y - X)U_2 + XU_5$	$1 - Y$
3	$(1 + X - Y)U_0 + YU_2 - XU_3$	$1 + X - Y$
4	$(1 + X)U_0 + (Y - X)U_3 - YU_6$	$1 + X$
5	$(1 + X)U_0 + (X - Y)U_4 + XU_6$	$1 + Y$
6	$(1 - X + Y)U_0 + XU_1 - YU_4$	$1 - X + Y$

Note that the test function $\phi(x, y)$ takes the value unity at the origin and vanishes

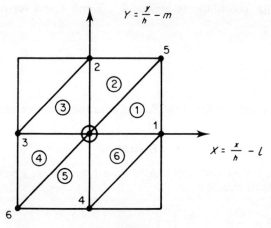

Figure 7

on the boundary of the union of the six elements. The integrals appearing in (34) are written as the sum of integrals over each of the elements and the individual contributions are tabulated:

Element number	$\iint \left(\dfrac{\partial U}{\partial x} \dfrac{\partial \phi}{\partial x} + \dfrac{\partial U}{\partial y} \dfrac{\partial \phi}{\partial y} \right) dx\, dy$	$\iint \phi\, dx\, dy$
1	$\frac{1}{2}(U_0 - U_1)$	$\frac{1}{6} h^2$
2	$\frac{1}{2}(U_0 - U_2)$	$\frac{1}{6} h^2$
3	$\frac{1}{2}(2U_0 - U_3 - U_2)$	$\frac{1}{6} h^2$
4	$\frac{1}{2}(U_0 - U_3)$	$\frac{1}{6} h^2$
5	$\frac{1}{2}(U_0 - U_4)$	$\frac{1}{6} h^2$
6	$\frac{1}{2}(2U_0 - U_4 - U_1)$	$\frac{1}{6} h^2$

Summing these contributions we obtain

$$4U_0 - \sum_{i=1}^{4} U_i = h^2 \tag{36}$$

for the difference replacement of the Poisson equation at the internal grid point.

5.6 Semi-discrete Galerkin methods

We now turn to time dependent partial differential equations and discretize the spatial part of the equation by the Galerkin method. This leads to a *system of ordinary differential equations in time.*

As an illustration consider the parabolic equation

$$\frac{\partial u(\mathbf{x},t)}{\partial t} + Au(\mathbf{x}, t) = 0 \qquad [(\mathbf{x}, t) \in R \times (t_0, t_1)] \qquad (37)$$

where in two dimensions, A is a second-order differential operator such as

$$A = -\left(\frac{\partial^2}{\partial x^2} + \frac{\partial^2}{\partial y^2}\right).$$

A solution of (37) is required subject to the initial condition

$$u(\mathbf{x}, t_0) = u_0(\mathbf{x}) \qquad (\mathbf{x} \in R)$$

and the boundary condition

$$u(\mathbf{x}, t) = 0 \qquad [(\mathbf{x}, t) \in \partial R \times (t_0, t_1)].$$

Using the notation introduced at the beginning of this chapter, we can write the *weak* form of the problem as

$$\left(\frac{\partial u}{\partial t}, v\right) + a(u, v) = 0 \qquad (\forall\, v(\mathbf{x}) \in H_0^1(R)) \qquad (38)$$

subject to the initial condition

$$(u, v)_{t=t_0} = (u_0, v) \qquad (\forall\, v(\mathbf{x}) \in H_0^1(R)).$$

Again we consider S^{M+1} to be a finite dimensional subspace of H_0^1, and the Galerkin approximation $U(\mathbf{x}, t)$ to the weak solution $u(\mathbf{x}, t)$ is written in the form

$$U(\mathbf{x}, t) = \sum_{i=0}^{M} U_i(t)\phi_i(\mathbf{x}) \qquad (39)$$

where the basis functions $\phi_i(\mathbf{x})$ $(i = 0, 1, \ldots, M)$ span the space S^{M+1}. From (38) and (39) we can now define the semi-discrete Galerkin approximation to be the element $U(\mathbf{x}, t)$ which satisfies

$$\left(\frac{\partial U}{\partial t}, \phi_j\right) + a(U, \phi_j) = 0 \qquad (40)$$

subject to

$$(U, \phi_j)_{t=t_0} = (u_0, \phi_j)$$

$\forall\, \phi_j \in S^{M+1}$. The equations (40) can be rewritten in the form

$$\sum_{i=1}^{M-1} [\dot{U}_i\, (\phi_i, \phi_j) + U_i\, a(\phi_i, \phi_j)] = 0 \qquad (j = 1, 2, \ldots, M\text{-}1), \qquad (41)$$

where a dot denotes differentiation with respect to t. The homogeneous boundary conditions have been incorporated into (41) and the initial condition becomes

$$\sum_{i=1}^{M-1} U_i(t_0)\,(\phi_i, \phi_j) = (u_0, \phi_j) \qquad (j = 1, 2, \ldots, M-1).$$

Example 5. $A \equiv \dfrac{\partial^2}{\partial x^2}$.

Equation (41) becomes

$$\sum_{i=1}^{M-1} [\dot{U}_i\,(\phi_i, \phi_j) + U_i(\phi_i', \phi_j')] = 0 \qquad (j = 1, 2, \ldots, M-1) \qquad (42)$$

where a prime denotes differentiation with respect to x. The basis functions ϕ_i $(i = 1, 2, \ldots, M-1)$ are chosen to be

(a) *Linear 'hat' functions* (see figure 1). Referring to Example 5 in the section on two-point boundary value problems, we see that $U_i(\phi_i', \phi_j')$ has already been evaluated. The quantity $\dot{U}_i\,(\phi_i, \phi_j)$ at the standard position leads to

$$[- \dot{U}_{-1} \int_{-1}^{0} X(1 + X)\mathrm{d}X + \dot{U}_0\,(\int_{-1}^{0}(1 + X)^2\,\mathrm{d}X +$$

$$\int_{0}^{1} (1 - X)^2\,\mathrm{d}X) + \dot{U}_{+1}\;\int_{0}^{1} X(1 - X)\mathrm{d}X]h^3$$

which on evaluation gives

$$\tfrac{1}{6}h\,(\dot{U}_{+1} + 4\dot{U}_0 + \dot{U}_{-1}).$$

Thus (42) leads to the system of equations

$$h^2 M\dot{\mathbf{U}} = S\mathbf{U} \qquad (43)$$

where $\mathbf{U} = (U_1, U_2, \ldots, U_{M-1})^T$, and the matrices M and S are given by

$$M = \tfrac{1}{6}\begin{bmatrix} 4 & 1 & & & & \\ 1 & 4 & 1 & & & \\ & \cdot & \cdot & \cdot & & \\ & & \cdot & \cdot & \cdot & \\ & & & \cdot & \cdot & 1 \\ & & & & 1 & 4 \end{bmatrix}, \text{ and } S = \begin{bmatrix} -2 & 1 & & & & \\ 1 & -2 & 1 & & & \\ & \cdot & \cdot & \cdot & & \\ & & \cdot & \cdot & \cdot & \\ & & & \cdot & \cdot & 1 \\ & & & & 1 & -2 \end{bmatrix}$$

Exercise

7. If $A \equiv - (\partial^2/\partial x^2) + K(\partial/\partial x)$, K a constant (> 0), show that the semi-discrete Galerkin method with linear basis functions leads to (43) where the matrix M is as above and

$$S = \begin{bmatrix} -2 & 1-L & & & & \\ 1+L & -2 & 1-L & & & \\ & \cdot & \cdot & \cdot & & \\ & & \cdot & \cdot & \cdot & \\ & & & \cdot & \cdot & 1-L \\ & & & 1+L & -2 \end{bmatrix}$$

with $L = \tfrac{1}{2} hK$.

(b) *Quadratic functions* (see figure 2). The quantity $U_i(\phi_i', \phi_j')$ at the standard position has already been evaluated at both integer and half-integer nodes. The quantity $\dot{U}_i(\phi_i, \phi_j)$ at an integer node leads to

$$U_{-1} \int_{-1}^{0} (X + 2X^2)(1 + 3X + 2X^2)dX + U_{-1/2} \int_{-1}^{0} 4(X + X^2)(1 + 3X + 2X^2)dX$$

$$+ \dot{U}_0 [\int_{-1}^{0} (1 + 3X + 2X^2)^2 \, dX + \int_{0}^{1} (1 - 3X + 2X^2)^2 \, dX]$$

$$+ \dot{U}_{+1/2} \int_{0}^{1} 4(X - X^2)(1 - 3X + 2X^2)dX + \dot{U}_{+1} \int_{0}^{1} (-X + 2X^2)(1 - 3X + 2X^2)dX$$

which on evaluation gives

$$\tfrac{1}{30} h \, (-U_{+1} + 2U_{+1/2} + 8U_0 + 2U_{-1/2} - U_{-1}).$$

At the half-integer node we have

$$\dot{U}_0 \int_{0}^{1} 4(1 - 3X + 2X^2)(X - X^2)dX + \dot{U}_{1/2} \int_{0}^{1} 16(X - X^2)^2 \, dX +$$

$$\int_{0}^{1} 4(-X + 2X^2)(X - X^2)dX$$

which gives

$$\tfrac{1}{15} h \, (U_{+1} + 8U_{+1/2} + U_0).$$

This time, the system of equations (43) has

$$\mathbf{U} = (U_{1/2}, U_1, \ldots, U_{M-1}, U_{M-1/2})^T,$$

$$M = \tfrac{1}{10}\begin{bmatrix} 8 & 1 \\ 2 & 8 & 2 & -1 \\ & 1 & 8 & 1 \\ & -1 & 2 & 8 & 2 & -1 \\ & & & \cdot & \cdot & \cdot & \cdot & \cdot \\ & & & & \cdot & \cdot & \cdot & \cdot & \cdot \\ & & & & & \cdot & \cdot & \cdot & \cdot & \cdot \\ & & & & & -1 & 2 & 8 & 2 & -1 \\ & & & & & & 1 & 8 & 1 \\ & & & & & & -1 & 2 & 8 & 2 \\ & & & & & & & & 1 & 8 \end{bmatrix}$$

$$\text{and } S = \begin{bmatrix} -8 & 4 \\ 8 & -14 & 8 & -1 \\ & 4 & -8 & 4 \\ & -1 & 8 & -14 & 8 & -1 \\ & & \cdot & \cdot & \cdot & \cdot & \cdot \\ & & & \cdot & \cdot & \cdot & \cdot & \cdot \\ & & & & -1 & 8 & -14 & 8 & -1 \\ & & & & & 4 & -8 & 4 \\ & & & & & -1 & 8 & -14 & 8 \\ & & & & & & & 4 & -8 \end{bmatrix}$$

Exercise

8. If $A \equiv \partial/\partial x$, show that the semi-discrete Galerkin method with quadratic basis functions leads to

$$hM\dot{\mathbf{U}} = S\mathbf{U}$$

where M is as above and

$$S = \begin{bmatrix} 0 & -1 \\ 2 & 0 & -2 & \tfrac{1}{2} \\ & 1 & 0 & -1 \\ & -\tfrac{1}{2} & 2 & 0 & -2 & \tfrac{1}{2} \\ & & \cdot & \cdot & \cdot & \cdot & \cdot \\ & & & \cdot & \cdot & \cdot & \cdot & \cdot \\ & & & & -\tfrac{1}{2} & 2 & 0 & -2 & \tfrac{1}{2} \\ & & & & & 1 & 0 & -1 \\ & & & & & -\tfrac{1}{2} & 2 & 0 & -2 \\ & & & & & & & 1 & 0 \end{bmatrix}$$

5.7 Discretization in time

We now look at the time discretization of the system of ordinary differential equations (41) by *finite difference methods*. From the examples it appears that (41) reduces to the form

$$h^{\alpha}M\dot{\mathbf{U}} = S\mathbf{U} \tag{44}$$

where $\alpha = 1,2$ in the hyperbolic and parabolic cases, and the matrices M and S are the mass and stiffness matrices respectively. Comprehensive studies of finite difference methods for the solution of systems like (44) can be found in Gear (1971) and Lambert (1972). In the time stepping methods to follow it is assumed that M and S are constant matrices. The extension to variable coefficients is straightforward.

Explicit difference methods are based on the discretization

$$\frac{h^\alpha}{k} M(\mathbf{U}^{n+1} - \mathbf{U}^n) = S\mathbf{U}^n,$$

where the equally spaced time intervals are at $t = nk, n = 0, 1, 2, \ldots$ .This leads to the *consistent mass* formulation

$$M\mathbf{U}^{n+1} = (M + \frac{k}{h^\alpha} S)\mathbf{U}^n. \tag{45a}$$

It is common practice, particularly with engineers, to replace the mass matrix M by the unit matrix I. This is known as *mass lumping*, and in (45) we can have either partial mass lumping which leads to

$$\mathbf{U}^{n+1} = (M + \frac{k}{h^\alpha} S)\mathbf{U}^n \tag{45b}$$

or complete mass lumping which leads to

$$\mathbf{U}^{n+1} = (I + \frac{k}{h^\alpha} S)\mathbf{U}^n \tag{45c}$$

It is interesting to carry out von Neumann stability analyses for the three methods above. For the parabolic equation $u_t = u_{xx}$ using linear basis functions for the space discretization we get the stability ranges $0 < r (= k/h^2) \leqslant \frac{1}{6}, \frac{1}{3}, \frac{1}{2}$ respectively. For the hyperbolic equation $u_t + u_x = 0$ with linear basis functions we get unconditional instability for methods (45a) and (45c) and the stability range $0 < p(= k/h) \leqslant 1/\sqrt{3}$ for method (45b). It appears therefore that mass lumping may improve the stability of a method. Its effect on *accuracy*, however, is more difficult to access and will not be discussed here (see Gresho et al. (1978)).

Implicit difference methods are usually based on the formula

$$\frac{h^\alpha}{k} M(\mathbf{U}^{n+1} - \mathbf{U}^n) = S[\theta\mathbf{U}^{n+1} + (1-\theta)\mathbf{U}^n] \quad (\tfrac{1}{2} < \theta \leqslant 1),$$

which leads to

$$(M - \frac{k}{h^\alpha}\theta S)\mathbf{U}^{n+1} = [M + \frac{k}{h^\alpha}(1-\theta) S]\mathbf{U}^n. \tag{46}$$

Mass lumping is not considered since it neither eases the work load nor improves the stability, method (46) being unconditionally stable. Common values for the parameter are $\theta = \frac{1}{2}$ (Crank–Nicolson) and $\theta = 1$ (backward difference).

5.8 Non-linear problems

In applications, *non-linear* problems are most important. However they can only be considered individually and as an example of a non-linear equation we consider the time-dependent Burgers' equation, which arises in model studies of turbulence and shock wave theory. It is the non-linear parabolic equation

$$\frac{\partial u}{\partial t} = \epsilon \frac{\partial^2 u}{\partial x^2} - u \frac{\partial u}{\partial x} \quad (0 < x < 1, t > 0) \tag{47}$$

with the initial condition

$$u(x, 0) = u_0(x) \quad (0 < x < 1)$$

and the homogeneous boundary conditions

$$u(0, t) = u(1, t) = 0 \quad (t \geqslant 0),$$

where ϵ (> 0) is the coefficient of kinematic viscosity. The mathematical properties of (47) have been studied by Cole (1951), and in the theory of shocks, (47) provides a model where the transport of momentum of the fluid, $u_t + uu_x$, is dissipated by the viscous term ϵu_{xx}.

The weak form of the problem is

$$(\frac{\partial u}{\partial t}, v) + \epsilon(\frac{\partial u}{\partial x}, \frac{\partial v}{\partial x}) + (u\frac{\partial u}{\partial x}, v) = 0 \quad [\forall\, v\,(x) \in H_0^1([0, 1])] \tag{48}$$

subject to the initial condition

$$(u, v)_{t=0} = (u_0, v) \quad [\forall\, v(x) \in H_0^1([0, 1])].$$

As in the linear case, the Galerkin approximation given by (39) satisfies

$$(\frac{\partial U}{\partial t}, \psi_j) + \epsilon(\frac{\partial U}{\partial x}, \frac{\partial \psi_j}{\partial x}) + (U\frac{\partial U}{\partial x}, \psi_j) = 0 \tag{49}$$

subject to

$$(U, \psi_j)_{t=0} = (u_0, \psi_j)$$

$\forall\, \psi_j \in S^{M+1}$. The equations (49) can be rewritten in the form

$$\sum_{i=1}^{M-1} [\dot{U}_i(\phi_i, \psi_j) + \epsilon U_i(\phi_i', \psi_j')] + (\sum_{i=1}^{M-1} U_i\phi_i \sum_{i=1}^{M-1} U_i\phi_i', \psi_j) = 0$$

$$(j = 1, 2, \ldots, (M-1)). \qquad (50)$$

With linear trial functions and the perturbed test functions defined by (25), equation (50) becomes

$$\tfrac{1}{12}h^2 [(2 + 3\alpha)\dot{U}_{j-1} + 8\dot{U}_j + (2 - 3\alpha)\dot{U}_{j+1}] =$$

$$\epsilon [U_{j-1} - 2U_j + U_{j+1}] - \tfrac{1}{12}h[2(U_{j-1} + U_j + U_{j+1})(U_{j+1} - U_{j-1})$$

$$- 3\alpha(U_{j-1}^2 - 2U_j^2 + U_{j+1}^2)] \qquad (51)$$

for $j = 1, 2, \ldots, M - 1$, where α is an upwinding parameter which is at our disposal. These equations together with the boundary conditions $U_0 = U_M = 0$ can be assembled into a matrix system such as (44) in which the matrix S now depends on U. When this system is advanced in time by the implicit difference method given by (46), a set of non-linear algebraic equations is obtained which must be solved at each time level. An iterative process must be used for this purpose and perhaps the Newton–Raphson method is the most suitable (see for instance Isaacson and Keller (1967)).

To demonstrate the need for upwinding at small values of ϵ, we consider the Galerkin solution of (47) with $\epsilon = 10^{-4}$ and the initial condition $u(x, 0) = \sin \pi x$.

The numerical results at $t = 1$ using the Crank–Nicolson method with a time increment $k = 0.001$ and $h = \tfrac{1}{18}$ are shown in table 1 both for no upwinding ($\alpha = 0$) and upwinding using the value of α presented by (29). We interpret this latter value by allowing α to vary from one mesh point to another so that, at the jth internal node, $\alpha = \alpha_j$ where

$$\alpha_j = \coth(hU_j^n/2\epsilon) - 2\epsilon/(hU_j^n).$$

Also shown in table 1 are the results obtained with quadratic basis functions and the upwinded quadratic functions described in section 5.4 using the values $\alpha_1 = 8$, $\alpha_2 = 0$ and $\alpha_3 = -2$. The value $h = \tfrac{1}{9}$ was used in these calculations so that the total number of internal nodes was the same as for linear trial functions. The results at integer nodes are underlined for emphasis.

The accurate solution quoted in table 1 was computed by the Galerkin method with fully upwinded cubic functions and a particularly small value of h. We are indebted to I. Christie for providing the numerical results and many of the details of the methods described in this section.

We now take the opportunity to describe two finite difference methods for solving equation (47).

Table 1

Node number	Accurate solution	Linear elements		Quadratic elements	
		No upwinding	Upwinding	No upwinding	Upwinding
0	0.0	0.0	0.0	0.0	0.0
1	0.0422	0.108	0.0421	0.091	0.0422
2	0.0843	−0.113	0.0842	−0.218	0.0843
3	0.1263	0.405	0.1263	0.246	0.1264
4	0.1684	−0.443	0.1682	−0.374	0.1683
5	0.2103	0.713	0.2103	0.287	0.2103
6	0.2522	−0.936	0.2518	−0.423	0.2521
7	0.2939	1.093	0.2942	−0.027	0.2939
8	0.3355	−1.401	0.3344	−1.710	0.3355
9	0.3769	1.607	0.3784	1.310	0.3769
10	0.4182	−0.369	0.4148	0.108	0.4182
11	0.4592	1.306	0.4650	0.935	0.4592
12	0.5000	−0.384	0.4879	−0.305	0.4999
13	0.5404	1.419	0.5627	1.057	0.5404
14	0.5806	−0.216	0.5357	−0.225	0.5805
15	0.6203	1.399	0.7041	1.056	0.6093
16	0.6596	−0.115	0.4861	−0.109	0.6166
17	0.6983	1.466	1.0053	1.115	0.5829
18	0.0	0.0	0.0	0.0	0.0

Compact differencing technique

By approximating the spatial derivatives in (47) independently, high order difference equations can be derived which retain the tridiagonal structure of the coefficient matrix. Denoting by F and S the approximations to $\partial u/\partial x$ and $\partial^2 u/\partial x^2$ respectively and using the expansions

$$\frac{\partial u}{\partial x} = \frac{1}{h}(1 + \tfrac{1}{6}\delta_x^2)^{-1}H_x u + 0(h^4)$$

[where $H_x u(x, t) = u(x + h, t) - u(x - h, t)$] and

$$\frac{\partial^2 u}{\partial x^2} = \frac{1}{h^2}(1 + \tfrac{1}{12}\delta_x^2)^{-1}\delta_x^2 u + 0(h^4)$$

we obtain, at the node (nk, mh)

$$F_{m+1}^n + 4F_m^n + F_{m-1}^n = \frac{3}{h}(U_{m+1}^n - U_{m-1}^n) \tag{52}$$

and

$$S_{m+1}^n + 10S_m^n + S_{m-1}^n = \frac{12}{h^2}(U_{m+1}^n - 2U_m^n + U_{m-1}^n) \tag{53}$$

for $m = 1, 2, \ldots, M - 1$. To complete these systems, additional equations are required at $m = 0$ and $m = M$. By means of Taylor expansions, it can be shown that

$$F_0^n + 6F_1^n + 3F_2^n = \frac{1}{3h}(U_3^n + 18U_2^n - 9U_1^n - 10U_0^n) \tag{54}$$

and

$$F_M^n + 6F_{M-1}^n + 3F_{M-2}^n = \frac{1}{3h}(U_{M-3}^n + 18U_{M-2}^n - 9U_{M-1}^n - 10U_M^n) \tag{55}$$

are $0(h^5)$ approximations to the equation $f = \partial u/\partial x$ at $m = 0$ and $m = M$ respectively. Since $S = \partial^2 u/\partial x^2 = \partial f/\partial x$, boundary equations for S can be obtained from (54) and (55) by replacing F by S and U by F. When these supplementary equations are used to eliminate F_0^n, F_M^n, S_0^n and S_M^n from (52) and (53), two tridiagonal systems are obtained which enable F and S to be economically computed from U.

With an explicit finite difference approximation of the time derivative, equation (47) is replaced by

$$(U_m^{n+1} - U_m^n)/k = \epsilon S_m^n - U_m^n F_m^n \quad (m = 1, 2, \ldots, M - 1).$$

The results obtained with this method when $k = 0.001$ and $h = \frac{1}{18}$ are shown in table 2 and should be compared with the corresponding results for the Galerkin methods presented in table 1.

Table 2

Node number	Accurate solution	Compact differencing	Splitting Method (i) $s = 6, k = 0.05$	(ii) $s = 0.5, k = 0.001$
0	0.0	0.0	0.0	0.0
1	0.0422	0.0501	0.0416	0.0422
2	0.0843	0.0753	0.0833	0.0845
3	0.1263	0.1471	0.1249	0.1267
4	0.1684	0.1359	0.1664	0.1688
5	0.2103	0.2611	0.2080	0.2108
6	0.2522	0.2091	0.2495	0.2528
7	0.2939	0.3340	0.2909	0.2946
8	0.3355	0.3048	0.3323	0.3362
9	0.3769	0.4173	0.3731	0.3782
10	0.4182	0.3741	0.4164	0.4176
11	0.4592	0.5059	0.4509	0.4652
12	0.5000	0.4634	0.5059	0.4886
13	0.5404	0.5808	0.5262	0.5680
14	0.5806	0.5369	0.5813	0.5305
15	0.6203	0.6671	0.6284	0.7269
16	0.6596	0.6201	0.6333	0.4589
17	0.6983	0.7410	0.7397	1.1477
18	0.0	0.0	0.0	0.0

The compact differencing technique is described fully by Collatz (1960) (under the name 'Mehrstellenverfahren') and, more recently, by Hirsh (1975) and Ciment et al. (1978). The device of introducing the new dependent variable f ($= \partial u / \partial x$) is also adopted by Cullen (1974) within a Galerkin finite element framework.

A splitting method

In contrast to section 4.7 where the differential equation is split according to its derivatives in the x and y directions, we now describe a splitting of (47) into its linear and non-linear components. This scheme was communicated to us by J. Ll. Morris (1978).

Replacing the spatial derivatives in (47) by the usual central difference approximations leads to the system of ordinary differential equations

$$\dot{U}_m(t) = \frac{\epsilon}{h^2} \, \delta_x^2 U_m \; - \; \frac{1}{2h} \, U_m H_x U_m \quad (m = 1, 2, \dots, M-1). \tag{56}$$

Thus, proceeding formally,

$$U_m \, (t + k) = \exp \, [k(L_1 + L_2)] \, U_m(t) \tag{57}$$

where the difference operators L_1 and L_2 are defined by

$$L_1 = \frac{\epsilon}{h^2} \, \delta_x^2 \text{ and } L_2 = - \frac{1}{2h} \, U_m H_x.$$

For $t = nk$ equations (57) can be rearranged to give

$$U_m^{n+1} = \exp(\tfrac{1}{2}kL_2) \exp(\tfrac{1}{2}kL_1) \exp(\tfrac{1}{2}kL_1) \exp(\tfrac{1}{2}kL_2) U_m^n \tag{58}$$

where the errors due to the non-commutativity of the exponential operators have been ignored. For computational purposes equation (58) is expressed in terms of intermediate quantities $U_m^{(1)}$, $U_m^{(2)}$ and $U_m^{(3)}$ as

$$U_m^{(1)} = \exp(\tfrac{1}{2}kL_2) U_m^n \tag{59a}$$

$$U_m^{(2)} = \exp(\tfrac{1}{2}kL_1) U_m^{(1)} \tag{59b}$$

$$U_m^{(3)} = \exp(\tfrac{1}{2}kL_1) U_m^{(2)} \tag{59c}$$

$$U_m^{n+1} = \exp(\tfrac{1}{2}kL_2) U_m^{(3)}. \tag{59d}$$

Equation (59a) is solved using the McGuire and Morris generalization of the two-step Lax–Wendroff formula (see equation (21), Chapter 4 with $p = \tfrac{1}{2}k/h$) so that

$$\bar{U}^{(1)}_{m+1/2} = \mu_x U^n_{m+1/2} - sp\, \mu_x U^n_{m+1/2}\, \delta_x U^n_{m+1/2} \tag{60a}$$

and

$$U^{(1)}_m = U^n_m - \tfrac{1}{2}p\,[(1 - \tfrac{1}{2}s)\,(U^n_{m+1} + U^n_{m-1})H_x\, U^n_m + \frac{1}{s}\, \mu_x\, \bar{U}^{(1)}_m)\, \delta_x\, \bar{U}^{(1)}_m] \tag{60b}$$

where $p = \tfrac{1}{2}k/h$. A similar pair of equations can be derived from (59d) to compute U^{n+1} from $U^{(3)}$.

The Crank–Nicolson method is used for the stages involving the linear operator L_1. This is equivalent to the approximation $\exp(z) \approx (1 + \tfrac{1}{2}z)/(1 - \tfrac{1}{2}z)$ and leads to

$$[1 - er\delta^2_x]\, U^{(2)}_m = [1 + er\delta^2_x]\, U^{(1)}_m \qquad (r = \tfrac{1}{2}k/h^2),$$

for equation (59b) with a similar expression for (59c).

The derivation of suitable boundary conditions for the intermediate variable poses considerable difficulties in general but, when U is subject to homogeneous Dirichlet conditions, these can also be used for $U^{(1)}$, $U^{(2)}$ and $U^{(3)}$. A certain amount of experimentation is required to determine the best combination of the parameters p and s in (60). In the results shown in table 2 these were found to be $k = 0.05$ and $s = 6$ and show a marked improvement over the Lax–Wendroff method ($k = 0.001, s = \tfrac{1}{2}$).

Chapter 6

Applications

6.1 Re-entrant corners and boundary singularities

The boundary ∂R of a region R is said to have a re-entrant corner at P if the boundary subtends an internal angle $\alpha\pi$ at P with $1 < \alpha \leqslant 2$. The boundary in the neighbourhood of P is shown in figure 1 where we have assumed that the edges AP

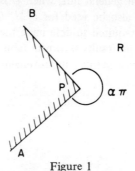

Figure 1

and PB are both linear. Adopting P as the origin of co-ordinates and then making a transformation to polar co-ordinates (r, θ) where

$$x = r\cos\theta, \quad y = r\sin\theta \tag{1}$$

(arranged so that $\theta = 0$ and $\theta = \alpha\pi$ correspond to the boundary arcs AP and PB respectively) it is possible to describe the solution of the governing differential equation in the neighbourhood of P in terms of an infinite series. For example, the solution of Laplace's equation may be written as

$$u = \sum_{n=1}^{\infty} a_n r^{n/\alpha} \sin(n\theta/\alpha) \quad \text{if} \quad u = 0 \text{ on AP and BP} \tag{2}$$

or as

$$u = \sum_{n=1}^{\infty} b_n r^{(n+\frac{1}{2})/\alpha} \sin(n + \frac{1}{2})\theta/\alpha \quad \text{if} \quad u = 0 \text{ on AP and } \frac{\partial u}{\partial \theta} = 0 \text{ on BP.} \tag{3}$$

The constants a_n and b_n $(n \geqslant 1)$ appearing in (2) and (3) may be determined approximately from boundary conditions not already satisfied (as in the work of Fox, Henrici and Moler (1967)) but we shall consider them to be unknown. Using the expressions

$$\frac{\partial u}{\partial x} = \cos \theta \, \frac{\partial u}{\partial r} - \frac{\sin \theta}{r} \frac{\partial u}{\partial \theta} \,, \quad \frac{\partial u}{\partial y} = \sin \theta \, \frac{\partial u}{\partial r} + \frac{\cos \theta}{r} \frac{\partial u}{\partial \theta} \,,$$

we find from equation (2) that

$$\frac{\partial u}{\partial x} = -\frac{a_1}{r^\beta} \sin \beta\theta + 0(r^{1-\beta})$$

and

$$\frac{\partial u}{\partial y} = \frac{a_1}{r^\beta} \cos \beta\theta + 0(r^{1-\beta})$$

where $\beta = 1 - 1/\alpha$ with similar expressions derived from (3). Consequently, if $\alpha > 1$, both $\partial u/\partial x$ and $\partial u/\partial y$ tend to infinity as $r \to 0$. That is to say, there is a *boundary singularity* at P $(r = 0)$. As singularities must also be present in higher derivatives of the solution u, it brings into question the validity of finite difference approximations near P since these depend for their success on the fact that the truncation error, which usually depends on the fourth derivatives of u, should tend to zero as the mesh size $h \to 0$. The analysis of Bramble, Hubbard and Zlamal (1968) reveals that the solution of the standard finite difference equations continues to converge to the corresponding solution of the differential equation as $h \to 0$ even in the presence of boundary singularities but that the rate of convergence and consequently the accuracy for a finite value of h may be unacceptable. We shall now, by means of a model problem discussed by Griffiths (1977), describe a simple technique whereby the accuracy of finite difference equations may be considerably improved.

The problem we consider arises from the slow rotation of a disc of unit radius in an infinite expanse of viscous fluid. The disc is assumed to rotate with constant angular velocity about an axis through its centre and normal to its plane. From the solution to this problem it can be shown that the function $\Phi(x, y)$ defined by

$$\Phi(x, y) = x(1 + f(\xi)), \qquad x \geqslant 0, \tag{4}$$

where

$$f(\xi) = 2\left[\xi/(1 + \xi^2) - \text{arc cot } \xi\right] / \pi$$

and ξ is the positive real root of the equation

$$\xi^4 - (x^2 + y^2 - 1)\xi^2 - y^2 = 0$$

satisfies the differential equation

$$Lu \equiv \frac{\partial^2 u}{\partial x^2} + \frac{1}{x}\frac{\partial u}{\partial x} + \frac{\partial^2 u}{\partial y^2} - \frac{u}{x^2} = 0, \qquad x > 0 \tag{5}$$

together with the conditions

$$u(0, y) = 0 \quad \text{for all } y$$

$$u(x, 0) = 0 \quad 0 \leqslant x \leqslant 1. \tag{6}$$

To avoid the difficulties associated with problems defined on infinite domains we shall consider the numerical solution of (5) on the square $[0 < x < 2, -1 < y < 1]$ and impose the additional boundary conditions

$$u(2, y) = \Phi(2, y) \qquad (-1 < y < 1),$$

$$u(x, \pm 1) = \Phi(x, \pm 1) \qquad (0 < x < 2). \tag{7}$$

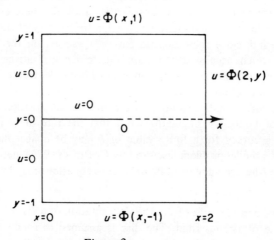

Figure 2

As we see from figure 2 the region of integration contains a 'crack' or re-entrant corner with interior angle 2π at O and consequently the solution is expected to have unbounded derivatives of all orders at this point. To be more explicit we make the transformation to polar co-ordinates (r, θ), centred on O, via

$$x = 1 + r \cos \theta, \qquad y = r \sin \theta. \tag{8}$$

A straightforward analysis then shows that in the neighbourhood of O, the solution of (5) has an asymptotic expansion of the form

$$u = A_0 (2r)^{1/2} \cos \tfrac{1}{2}\theta - \tfrac{1}{8}A_0(2r)^{3/2} \cos \tfrac{1}{2}\theta + A_1(2r)^{3/2} \cos \tfrac{3}{2}\theta + O(r^{5/2}). \tag{9}$$

Since the solution of the problem is known to be given by (4), the constants $A_0, A_1, \ldots$ can be determined by expressing (4) in polar co-ordinates and expanding as in (9) to give

$$A_0 = 4/\pi \approx 1.27324 \ldots, \qquad A_1 = (3\pi)^{-1} \approx 0.10610 \ldots \tag{10}$$

We emphasize that when dealing with general elliptic problems having boundary singularities, an expansion of the form (9) is usually possible but that the values of the coefficients appearing are not known. As we shall see below, the accuracy of numerical methods for the solution of (5) depends crucially on the degree to which these constants can be approximated.

As an indication of the general behaviour of the solution we note that $\Phi(x, y) \approx x$ on the boundary. Finally, in view of the symmetry about $y = 0$, we shall consider only the reduced domain $R = (0 < x < 2, 0 < y < 1)$ and impose the further condition

$$\frac{\partial u}{\partial y}(x, 0) = 0 \qquad (1 < x < 2). \tag{11}$$

We begin by imposing a regular mesh of side h on R and, at each internal node of this mesh, we replace the differential equation (5) by its usual centred difference approximation

$$\frac{1}{h^2} [x_{i+1/2}(U_{i+1,j} - U_{i,j}) - x_{i-1/2}(U_{i,j} - U_{i-1,j})] +$$

$$\frac{x_i}{h^2}(U_{i,j+1} - 2U_{i,j} + U_{i,j-1}) - \frac{1}{x_i} U_{i,j} = 0, \tag{12}$$

where $U_{i,j}$ denotes the finite difference approximation at $x_i = 1 + ih$, $y_j = jh$. Along $y = 0$ the boundary condition (11) is replaced by its simplest centred difference approximation.

The analysis of Bramble, Hubbard and Zlamal (1968) suggests that the error at a point distant r from the singularity behaves asymptotically as $h^{1/2-\epsilon}r^\epsilon$ as $h \to 0$, where $\epsilon > 0$. Thus, even though the classical truncation error estimates become unbounded in the neighbourhood of the singularity, the convergence of the method is assured.

To improve the accuracy of the method we follow the approach of Fox and Sankar. In our implementation the difference equations (12) are applied throughout the discretized region except at the 'inner' nodes I_1 and I_2 adjacent to the singularity (see figure 3). In view of (9), the solution at these nodes is assumed to be given by

$$u = a_0(2r)^{1/2} \cos \tfrac{1}{2}\theta \tag{13}$$

242

where the coefficient a_0 is obtained by matching at the 'outer' point O_1 to give

$$u = U_{2,0}(r/2h)^{1/2} \cos \tfrac{1}{2}\theta, \tag{14}$$

where, in the notation of (12) $U_{2,0}$ refers to the nodal value at O_1. Application of (14) at I_1 and I_2 provides values for $U_{1,0}$ and $U_{0,1}$ which are eliminated from (12). The resulting system is then solved for $U_{i,j}$.

Fox and Sankar (1969) point out that it is possible to obtain a heuristic error estimate if the nodal values at T_1 and T_2 using this method are compared with the values obtained by application of (14) directly at these points. The usefulness of such estimates can be seen from the results of table 4 in the next section.

Further terms from the asymptotic expansion (9) are easily incorporated. For example (13) may be replaced by

$$u = a_0(2r)^{1/2}(1 - \tfrac{1}{4}r) \cos \tfrac{1}{2}\theta + a_1(2r)^{3/2} \cos \tfrac{3}{2}\theta \tag{15}$$

where the parameters a_0 and a_1 are now obtained by matching at the two outer points O_1 and O_2 of figure 3. The application is then as above and error estimates can again be given.

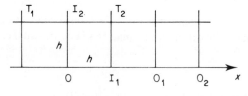

Figure 3

The choice of inner and outer nodes is obviously of some importance to the success of the method and, of the numerous possibilities that were tested, the arrangement described above was found to give the best results. It differs from that given by Fox and Sankar (1969) but this is to be expected since the problems considered have different symmetry properties about $y = 0$.

To assess the accuracy of the method for a range of mesh sizes we introduce the quantity

$$e(H) = \max_{(x,\, y)\, \in\, R_H} | U(x, y) - u(x, y)|, \tag{16}$$

where

$$R_H = [(pH, qH), \quad p, q \text{ integers}],$$

u and U denote respectively the exact and approximate solutions and $H(> 0)$ is taken as an integer multiple of h to ensure that $R_H \subset R_h$. The quantity $e(H)$ is a compromise between the overall maximum error in R and the tabulation of the error at a fixed set of points. The former can only be computed over points in R_h for finite difference methods whilst the latter can give a useful comparison of methods only if a sufficiently large number of points is taken.

The errors obtained when this method is applied for a range of values of h are shown in table 1. It is seen that the errors are an order of magnitude smaller than those obtained by ignoring the singularity. Further, there appears to be little difference between finite difference methods matching with one or two singular terms but further experiments show that two terms are superior for smaller values of h. The derivation of an error estimate was discussed earlier and the results of table 2 show that it is indeed a useful guide.

The treatment of boundary singularities in parabolic problems is discussed by Crank and Furzeland (1977) who also include a useful bibliography.

Table 1

	$1/h$	$e(1/16)$	$e(1/4)$	A_0	A_1
Exact values				1.27324	0.10610
Finite difference methods					
Ignoring	4		1.22(−1)		
singularity	8		5.93(−2)		
	16	7.97(−2)	2.76(−2)		
Matching with	4		6.89(−2)	1.2884	
one asymptotic	8		9.91(−3)	1.2697	
term (14)	16	8.17(−3)	2.16(−3)	1.2666	
Matching with	4		7.29(−2)	1.1891	0.2199
two asymptotic	8		1.33(−2)	1.2285	0.1978
terms (15)	16	7.34(−3)	1.87(−3)	1.2529	0.1754

Table 2
Error estimates for finite difference methods

	Finite difference methods matching with:			
	(a) one singular term (14)		(b) two singular terms (15)	
$1/h$	$e(h)$	Error estimate	$e(h)$	Error estimate
4	0.0689	0.0694	0.0729	0.0630
8	0.0213	0.0206	0.0243	0.0215
16	0.0082	0.0056	0.0073	0.0077

6.2 Incompressible viscous flow

In non-dimensional notation, the basic equations in the primitive variables u, v, p for *unsteady* incompressible viscous flow in two space dimensions are the Navier–Stokes equations

$$\frac{\partial u}{\partial t} = \frac{1}{Re} \nabla^2 u - u \frac{\partial u}{\partial x} - v \frac{\partial u}{\partial y} - \frac{\partial p}{\partial x}$$

$$\frac{\partial v}{\partial t} = \frac{1}{Re} \nabla^2 v - u \frac{\partial v}{\partial x} - v \frac{\partial v}{\partial y} - \frac{\partial p}{\partial y}$$

(17)

and the continuity equation

$$\frac{\partial u}{\partial x} + \frac{\partial v}{\partial y} = 0 .$$

(18)

The *conservation* form of (17), using (18), is

$$\frac{\partial u}{\partial t} = \frac{1}{Re} \nabla^2 u - \frac{\partial}{\partial x}(u^2) - \frac{\partial}{\partial y}(uv) - \frac{\partial p}{\partial x}$$

$$\frac{\partial v}{\partial t} = \frac{1}{Re} \nabla^2 v - \frac{\partial}{\partial x}(uv) - \frac{\partial}{\partial y}(v^2) - \frac{\partial p}{\partial y} .$$

(17a)

The pressure p is eliminated in (17) to give the vorticity transport equation (parabolic)

$$\frac{\partial \omega}{\partial t} = L\omega$$

where

$$L = \frac{1}{Re} \nabla^2 - \mathbf{q} . \nabla$$

(19)

together with the Poisson equation (elliptic)

$$\nabla^2 \psi = -\omega.$$

(20)

In the above we have

$$u = \frac{\partial \psi}{\partial y} , \quad v = -\frac{\partial \psi}{\partial x}, \quad \omega = \frac{\partial v}{\partial x} - \frac{\partial u}{\partial y} ,$$

where $\mathbf{q} = (u, v)$. Also if we use (18), equation (19) can be written in *conservation* form as

$$\frac{\partial \omega}{\partial t} = \frac{1}{Re} \nabla^2 \omega - \nabla . (\omega q) \tag{19a}$$

where the terms on the right-hand side are termed diffusion (conduction) and convection terms respectively. The non-dimensional notation requires the introduction of a standard velocity U and a standard length l, and so in the equations above x, y, t, u, v, ψ and ω represent the non-dimensional quantities x/l, y/l, $t/(l/U)$, u/U, v/U, ψ/Ul and $\omega/(U/l)$ respectively. Finally the important non-dimensional Reynolds number Re is given by

$$Re = \frac{Ul}{v} ,$$

where v is the coefficient of kinematic viscosity. The pressure p can be recovered from a solution based on (19) and (20) by solving the Poisson equation

$$\Delta^2 p = 2\left[\left(\frac{\partial^2 \psi}{\partial x^2}\right) \left(\frac{\partial^2 \psi}{\partial y^2}\right) - \left(\frac{\partial^2 \psi}{\partial x \partial y}\right)^2 \right] \tag{21}$$

The region to be examined in (x, y, t) space is covered by a rectilinear grid with grid spacings h and k in the space and time directions. Using the notation of Chapter 2 for a parabolic equation in two space dimensions, the vorticity transport equation (19) can be replaced by the *explicit* difference formula

$$\Omega_{l,m}^{n+1} = [1 + \frac{r}{Re}(\delta_x^2 + \delta_y^2) - \frac{k}{2h}(uH_x + vH_y)] \Omega_{l,m}^n , \tag{22}$$

where $\Omega_{l,m}^n$ is the vorticity at the grid point (lh, mh, nk), and $H_x = \Delta_x + \nabla_x$, $H_y = \Delta_y + \nabla_y$, with Δ and ∇ the forward and backward difference operators respectively. Notice in (22) that we have used second-order *central* difference operators to replace the first-order convection terms. The calculation starts at $t = 0$ where $\psi(u, v)$ and ω are known. From (22) with $n = 0$, we obtain the vorticity $\Omega(\omega)$ at $t = k$. The numerical solution of (20) gives ψ at $t = k$ and hence u and v. The process continues with the boundary values (if present) being included as the calculation proceeds in time.

We now apply the von Neumann stability criterion (see Chapter 2) to (22) with u and v assumed constant. The amplification factor is

$$\xi = [1 - \frac{4k}{h^2 Re}(\sin^2 \tfrac{1}{2}\beta h + \sin^2 \tfrac{1}{2}\gamma h)] - i\frac{k}{h}(u \sin \beta h + v \sin \gamma h). \tag{23}$$

The analysis of this expression proves to be intractable for general angles β, γ. Accordingly we look at the simplified case where

$$\gamma = \beta$$

and the von Neumann condition now gives

$$\xi^2 = 1 + (A^2 - 2B)s + (B^2 - A^2)s^2 \leqslant 1, \tag{24}$$

for $0 \leqslant s \leqslant 1$, where $A = 2k(u + v)/h$, $B = 8r/Re$, and $s = \sin^2 \tfrac{1}{2}\beta h$. Now $\xi^2 = 1$ at $s = 0$, and $\xi^2 = (1 - B)^2$ at $s = 1$, and so (24) is satisfied if

$$A < B \leqslant 2. \tag{25}$$

The latter inequality ensures that (24) is satisfied at $s = 1$ and the former that ξ^2 has a minimum in the range $0 < s < 1$. The conditions in (25) can be rewritten as

$$h Re (u + v) < 4, \tag{26a}$$

and

$$k \leqslant \tfrac{1}{4} h^2 Re, \tag{26b}$$

where $h Re (u + v)$ is the *cell Reynolds number*. The inequalities (26a) and (26b) will in general provide only necessary conditions for stability since they do not take account of boundary conditions.

We now return to (19), this time replacing the first-order convection terms by first-order difference replacements, *backward* or 'upwind' if u (or v) is positive and *forward* or 'downwind' if u (or v) is negative. If we assume u and v to be positive, (19) is replaced by

$$\Omega_{l,m}^{n+1} = [1 + \frac{r}{Re}(\delta_x^2 + \delta_y^2) - \frac{k}{h}(u \nabla_x + v \nabla_y)] \Omega_{l,m}^n, \tag{27}$$

for which the von Neumann amplification factor is

$$\xi = [1 - \frac{4k}{h^2 Re}(\sin^2 \tfrac{1}{2}\beta h + \sin^2 \tfrac{1}{2}\gamma h) - \frac{2k}{h}(u \sin^2 \tfrac{1}{2}\beta h + v \sin^2 \tfrac{1}{2}\gamma h)]$$

$$-i\frac{k}{h}(u \sin\beta h + v \sin \gamma h). \tag{28}$$

If

$$\gamma = \beta,$$

(28) gives

$$\xi^2 = 1 + [A^2 - 2(A + B)] s + (B^2 + 2AB)s^2 \tag{29}$$

for $0 \leqslant s \leqslant 1$. Now $\xi^2 = 1$ at $s = 0$ and $\xi^2 = (1 - A - B)^2$ at $s = 1$, and so

$$\xi^2 \leqslant 1$$

if

$$A + B \leqslant 2,$$

which leads to

$$k \leqslant \frac{h^2 Re}{h Re(u + v) + 4} \tag{30}$$

This corresponds to condition (26b) in the central difference case. Condition (26a) in the central difference case has no counterpart in the upwind stability analysis since the coefficient of s^2 in (29) is always positive.

Another explicit difference method for solving (19) can be obtained by following the locally one dimensional (L.O.D.) construction outlined in section 2.13. Using the Lax–Wendroff replacement for the convective part of the equation leads to

$$\Omega_{l,m}^{n+1/2} = (1 - \frac{kv}{2h} H_y + \tfrac{1}{2} \frac{k^2 v^2}{h^2} \delta_y^2 + \frac{1}{Re} \frac{k}{h^2} \delta_y^2)\Omega_{l,m}^n \tag{31a}$$

over the first half of the time step, and

$$\Omega_{l,m}^{n+1} = (1 - \frac{ku}{2h} H_x + \tfrac{1}{2} \frac{k^2 u^2}{h^2} \delta_x^2 + \frac{1}{Re} \frac{k}{h^2} \delta_x^2)\Omega_{l,m}^{n+1/2}, \tag{31b}$$

over the second half.

Exercise

1. Apply the von Neumann stability analysis to (31a) and, with the notation used in this section, show that

$$\xi^2 = 1 - Bs + 4[(\frac{k^2 v^2}{h^2} + \tfrac{1}{4}B)^2 - \frac{k^2 v^2}{h^2}]s^2.$$

Hence derive the stability condition

$$k < [(\frac{1}{Re^2} + v^2 h^2)^{1/2} - \frac{1}{Re}]/v^2.$$

We now turn to *implicit* difference methods for solving (19). Although these have the advantage of improved stability over explicit methods, they suffer from the disadvantage of requiring the solution of a set of equations at each time level. This problem is eased by using A.D.I. methods (section 2.12) which require the inversion of two tridiagonal matrices at each time step. For example, applying the Peaceman–Rachford A.D.I. method to the solution of (19), we get

$$[1 + \tfrac{1}{4} \frac{uk}{h} H_x - \tfrac{1}{2} \frac{k}{Re h^2} \delta_x^2]\Omega_{l,m}^{n+1/2} = [1 - \tfrac{1}{4} \frac{vk}{h} H_y + \tfrac{1}{2} \frac{k}{Re h^2} \delta_y^2]\Omega_{l,m}^n \tag{32}$$

$$[1 + \tfrac{1}{4} \frac{vk}{h} H_y - \tfrac{1}{2} \frac{k}{Re h^2} \delta_y^2]\Omega_{l,m}^{n+1} = [1 - \tfrac{1}{4} \frac{uk}{h} H_x + \tfrac{1}{2} \frac{k}{Re h^2} \delta_x^2]\Omega_{l,m}^{n+1/2}$$

Finally odd–even hopscotch described in section 2.14 is used by Jacobs (1971) to solve (19). The calculation starts at $t = nk$, where n, say is even, and the value of ω at the advanced time level is obtained in two consecutive sweeps:

(a) for all grid points with $(l + m + n)$ odd, calculate the value of ω at the advanced time step using

$$\Omega_{l,m}^{n+1} = (1 + kL_h)U_{l,m}^n \qquad (33a)$$

where L_h is a difference replacement of the operator L in (19), and
(b) for all grid points with $(l + m + n)$ even, calculate $\Omega_{l,m}^{n+1}$ using

$$(1-kL_h)\Omega_{l,m}^{n+1} = \Omega_{l,m}^n. \qquad (33b)$$

For difference operators of the types used in (22) and (27), the calculation based on (33b) is explicit. A stability analysis of odd–even hopscotch for solving (19) is very complicated. Interested readers are referred to Gourlay (1970) who shows the method to be unconditionally stable when upwind differencing is used for the convection term.

No mention has so far been made of boundary conditions or the shape of the region in the (x, y) plane. For explicit methods the shape of the region does not pose too great a problem, whereas for A.D.I. methods it is desirable that the boundary be of rectangular type. Boundary conditions, on the other hand, can present a considerable problem. The interested reader is referred to Roach (1972) (chapter 3C).

We now turn to the *steady* two-dimensional motion of an incompressible viscous fluid where, from (19), the basic equations are

$$\left(\frac{1}{Re}\nabla^2 - \mathbf{q}\cdot\nabla\right)\omega = 0, \qquad (34)$$

with

$$\omega = -\nabla^2\psi. \qquad (35)$$

The elliptic equation (34) in conjunction with (35) can be solved using finite differences either by an iterative method or by an unsteady method like (22) or (27) where the required solution is the limit of the unsteady solution as $t \to \infty$. It is often the case that an example of the latter corresponds to a particular iterative method for solving the elliptic equation. Another approach which is used frequently is to rewrite (34) and (35) as the fourth-order elliptic equation in the single variable ψ, viz.

$$\frac{1}{Re}\nabla^4\psi = \frac{\partial}{\partial x}\left(\frac{\partial \psi}{\partial y}\nabla^2\psi\right) - \frac{\partial}{\partial y}\left(\frac{\partial \psi}{\partial x}\nabla^2\psi\right), \qquad (36)$$

which of course for vanishing Reynolds number reduces to the *biharmonic* equation

$$\nabla^4 \psi = 0. \tag{37}$$

Finite difference methods for the solution of fourth-order elliptic equations can be found in Chapter 3.

Although we have concentrated in this section solely on the vorticity transport equation, we cannot ignore the numerical solution of the elliptic stream function equation (20). This requires solutions at every time step and so an efficient method of solution is desirable. A complete resumé of iterative and direct methods for the finite difference solution of second-order elliptic equations is given in Chapter 3.

So far the vorticity, stream function (ω, ψ) approach has dominated for the numerical solution of the Navier–Stokes equations. There are, however, many good finite difference methods based on equations (17) and (18). In particular, the TEACH programme of Imperial College, London, solves the steady state Navier–Stokes equations in primitive variables and conservation form over a rectangular grid by iterating over interlocking velocity and pressure grids. A generalization of TEACH from rectangular elements to isoparametric elements has been carried out by Wachspress (1977).

A number of novel ideas concerning finite difference approximations to (17) and (18) were introduced by Chorin (1969) but this work seems not to have attracted the attention that it deserves.

6.3 Inviscid compressible flow

In many problems in aerodynamics, significant parts of the flow field can be represented by the Eulerian equations of compressible (but otherwise idealized) fluid dynamics. In *two* space dimensions these can be expressed by the laws of *conservation* of mass, momentum, and energy respectively, viz.

$$\frac{\partial W}{\partial t} + \frac{\partial F}{\partial x} + \frac{\partial G}{\partial y} = 0, \tag{38}$$

where

$$W = \begin{bmatrix} \rho \\ \rho u \\ \rho v \\ e \end{bmatrix}, \quad F = \begin{bmatrix} \rho u \\ \rho u^2 + p \\ \rho u v \\ (e+p)u \end{bmatrix}, \text{ and } G = \begin{bmatrix} \rho v \\ \rho u v \\ \rho v^2 + p \\ (e+p)v \end{bmatrix}$$

with

$$e = \frac{p}{\gamma - 1} + \frac{1}{p} \rho(u^2 + v^2).$$

Along with these conservation laws we require the 'perfect gas' law

$$\frac{p}{\rho} = RT \tag{39}$$

and an entropy condition which prevents the occurrence of negative shocks.

Equation (38) represents a first-order *hyperbolic* system and many of the difference formulae given in Chapter 4 can be used for its numerical solution. The incorporation of boundary conditions into numerical solutions is a major problem and the interested reader is advised to consult the comprehensive treatise of Roache (1972) (chapter 5). We shall content ourselves with giving two finite difference schemes which are in common use for the numerical solution of (38).

(1) Two-step Lax–Wendroff

$$\bar{\mathbf{U}}_{l,m} = \tfrac{1}{4}(\mathbf{U}_{l+\frac{1}{2},m} + \mathbf{U}_{l-\frac{1}{2},m} + \mathbf{U}_{l,m+\frac{1}{2}} + \mathbf{U}_{l,m-\frac{1}{2}})^n$$

$$- \tfrac{1}{2}p(\mathbf{F}_{l+\frac{1}{2},m} - \mathbf{F}_{l-\frac{1}{2},m} + \mathbf{G}_{l,m+\frac{1}{2}} - \mathbf{G}_{l,m-\frac{1}{2}})^n \tag{40a}$$

$$\mathbf{U}_{l,m}^{n+1} = \mathbf{U}_{l,m}^n - p(\bar{\mathbf{F}}_{l+\frac{1}{2},m} - \bar{\mathbf{F}}_{l-\frac{1}{2},m} + \bar{\mathbf{G}}_{l,m+\frac{1}{2}} - \bar{\mathbf{G}}_{l,m-\frac{1}{2}}) \tag{40b}$$

where $p = k/h$. The first step (40a) is a *Lax* type step and the second step (40b) is a *leap-frog* type step. Both steps are based on central differences in space.

Exercise

2. For the scalar hyperbolic equation

$$\frac{\partial u}{\partial t} + \frac{\partial u}{\partial x} + \frac{\partial u}{\partial y} = 0 \,,$$

show, by elimination of the intermediate value, $\bar{U}_{l,m}$, that formula (40) reduces to the composite formula

$$U_{l,m}^{n+1} = U_{l,m}^n + \tfrac{1}{4}p(U_{l-1,m} + 2U_{l-\frac{1}{2},m-\frac{1}{2}} + U_{l,m-1} - U_{l+1,m} - 2U_{l+\frac{1}{2},m+\frac{1}{2}}$$

$$- U_{l,m+1})^n + \tfrac{1}{2}p^2(U_{l-1,m} + U_{l,m-1} + U_{l+1,m} + U_{l,m+1} + 2U_{l-\frac{1}{2},m-\frac{1}{2}}$$

$$+ 2U_{l+\frac{1}{2},m+\frac{1}{2}} - 2U_{l-\frac{1}{2},m+\frac{1}{2}} - 2U_{l+\frac{1}{2},m-\frac{1}{2}} - 4U_{l,m})^n .$$

Use Taylor expansions to show that this formula has second-order accuracy and so is comparable to a Lax–Wendroff formula.

(2) Two-step MacCormack (1969, 1971)

There are so many versions of this in two space dimensions that we shall give the method in one space dimension and indicate how the various two space dimensional versions may be obtained. It is

$$\bar{U}_m = U_m^n - p(F_{m+1}^n - F_m^n) \tag{41a}$$

$$U_m^{n+1} = \tfrac{1}{2}[U_m^n + \bar{U}_m - p(\bar{F}_m - \bar{F}_{m-1})] \tag{41b}$$

Here the first step is based on a *forward* difference in space and the second step on a *backward* difference (F.B.). Alternatively, we can have the backward difference in (41a) followed by the forward difference in (41b) (B.F.).

Exercise

3. Write out the backward—forward version of the two-step MacCormack scheme (B.F.) for the scalar hyperbolic equation

$$\frac{\partial u}{\partial t} + \frac{\partial u}{\partial x} = 0$$

and show by elimination of the intermediate value $\bar{U}_m$ that it reduces to the standard Lax—Wendroff formula.

In two space dimensions we can have variations of (F.B.) and (B.F.) in x and y leading to four possible schemes. Each pattern can also be applied cyclically over a number of time steps.

Example 1. Write down the two-step MacCormack scheme which uses (F.B.) in x and (B.F.) in y over a complete time step. Show that for the scalar equation

$$\frac{\partial u}{\partial t} + \frac{\partial u}{\partial x} + \frac{\partial u}{\partial y} = 0$$

this reduces to a Lax—Wendroff type scheme in two space dimensions.

The two-step MacCormack scheme is

$$\bar{U}_{l,m} = U_{l,m}^n - p(F_{l+1,m} - F_{l,m} + G_{l,m} - G_{l,m-1})^n \,,$$
$$U_{l,m}^{n+1} = \tfrac{1}{2}[U_{l,m}^n + \bar{U}_{l,m} - p(\bar{F}_{l,m} - \bar{F}_{l-1,m} + \bar{G}_{l,m+1} - \bar{G}_{l,m})]\,.$$

Rewriting this with F and G replaced by U followed by the elimination of $\bar{U}_{l,m}$ gives

$$U_{l,m}^{n+1} = U_{l,m}^n - \tfrac{1}{2}p(U_{l+1,m} - U_{l-1,m} + U_{l,m+1} - U_{l,m-1})^n$$
$$+ \tfrac{1}{2}p^2(U_{l+1,m+1} - 2U_{l,m} + U_{l-1,m-1})^n.$$

Expansion of the right-hand side by Taylor series leads to

$$u_{l,m}^{n+1} = [1 - k(\frac{\partial}{\partial x} + \frac{\partial}{\partial y}) + \tfrac{1}{2}k^2(\frac{\partial}{\partial x} + \frac{\partial}{\partial y})^2]u_{l,m}^n + \ldots$$

which is analogous to the derivation of the Lax—Wendroff method described in Chapter 4.

252

6.4 Other flows

Computation of *viscous compressible* flows based on the Navier–Stokes equations is not an exercise for the weak-willed. Interested readers are referred to an excellent treatise by Peyret and Viviand (1975). At the other end of the spectrum there are many problems in fluid mechanics where the relevant differential equations are gross simplifications of the full Navier–Stokes equations. Finite difference methods are used extensively in all areas which include (in no particular order)

  potential flow
  inviscid transonic flow
  boundary layer flow (laminar and turbulent)
  shock wave development (characteristics)
  flow behind shocks (non-isentropic).

The literature abounds with numerical work (often based on finite differences) on these topics.

6.5 Shallow water equations

The primitive equations describing incompressible inviscid flow with a free surface are called the shallow water equations. In two space dimensions the equations take the form

$$\frac{\partial u}{\partial t} + u\frac{\partial u}{\partial x} + v\frac{\partial u}{\partial y} + \frac{\partial \phi}{\partial x} - fv = 0$$

$$\frac{\partial v}{\partial t} + u\frac{\partial v}{\partial x} + v\frac{\partial v}{\partial y} + \frac{\partial \phi}{\partial y} - fu = 0 \qquad (42)$$

$$\frac{\partial \phi}{\partial t} + \frac{\partial}{\partial x}(\phi u) + \frac{\partial}{\partial y}(\phi v) = 0$$

where

  u, v velocity components in x- and y-directions.
  $\phi = gh$ geopotential.
  g acceleration due to gravity.
  $f = f_0 + \beta(y - \tfrac{1}{2}D)$ Coriolis parameter (f_0, β constants).
  h height of water surface.

The numerical solution of (42) is usually required in a rectangular region $0 \leqslant x \leqslant L$, $0 \leqslant y \leqslant D$ and for $t \geqslant 0$. The boundary conditions are *periodic* in the x-direction

$$\mathbf{w}(x, y, t) = \mathbf{w}(x + L, y, t) \qquad [\mathbf{w} = (u, v, \phi)^T],$$

and *rigid wall* in the y-direction

$$v(x, 0, t) = v(x, D, t) = 0, \tag{43}$$

and the initial condition is

$$\mathbf{w}(x, y, 0) = F(x, y).$$

With these boundary and initial conditions the energy

$$E = \tfrac{1}{2} \int_0^L \int_0^D (u^2 + v^2 + \phi) \frac{\phi}{g} \, dx \, dy$$

is independent of time, and this can be used as a check on approximate finite difference solutions.

The finite difference solution of (42) with (43) is usually required for large values of the time, and because the discretization error in time is often small compared to the discretization error in space it is convenient to take fairly large time steps. Due to restrictive stability conditions, this more or less rules out explicit difference approximations. Consequently we look at two difference schemes for solving (42), the first being partly implicit and the second fully implicit.

(i) *The partly implicit scheme.*

This is due to Navon (1977) and is written in the form

$$\phi_{l,m}^{n+1} - \phi_{l,m}^n + k\,A_\phi(U_{l,m}^n, V_{l,m}^n, \phi_{l,m}^{n+1}) = 0$$

$$U_{l,m}^{n+1} - U_{l,m}^n + k\,A_U(U_{l,m}^{n+1}, V_{l,m}^n, \phi_{l,m}^{n+1}) = 0 \tag{44}$$

$$V_{l,m}^{n+1} - V_{l,m}^n + k\,A_V(U_{l,m}^{n+1}, V_{l,m}^{n+1}, \phi_{l,m}^{n+1}) = 0$$

where, if we use second-order central differences in space

$$2h\,A_\phi = (U_{l,m}^n H_x + V_{l,m}^n H_y)\phi_{l,m}^{n+1} + \phi_{l,m}^{n+1}(H_x U_{l,m}^n + H_y V_{l,m}^n)$$

$$2h\,A_U = (U_{l,m}^{n+1} H_x + V_{l,m}^n H_y)U_{l,m}^{n+1} + H_x \phi_{l,m}^{n+1} - 2h f_{l,m} V_{l,m}^n \tag{45}$$

$$2h\,A_V = (U_{l,m}^{n+1} H_x + V_{l,m}^{n+1} H_y)V_{l,m}^{n+1} + H_y \phi_{l,m}^{n+1} - 2h f_{l,m} U_{l,m}^{n+1}.$$

The solution of (44) and (45) is achieved at each time step by solving a linear system for ϕ followed by two non-linear systems for U and V respectively. The non-linear systems can be *linearized* by replacing the terms $U_{l,m}^{n+1}H_xU_{l,m}^{n+1}$ and $V_{l,m}^{n+1} H_y V_{l,m}^{n+1}$ by linear approximations. For example

$$U_{l,m}^{n+1} U_{l+1,m}^{n+1} \approx U_{l,m}^n U_{l+1,m}^{n+1} + U_{l,m}^{n+1} U_{l+1,m}^n - U_{l,m}^n U_{l+1,m}^n$$

and similarly for $U_{l,m}^{n+1}$ $U_{l-1,m}^{n+1}$, $V_{l,m}^{n+1}$ $V_{l,m+1}^{n+1}$, and $V_{l,m}^{n+1}$ $V_{l,m-1}^{n+1}$. The local accuracy of this scheme is second-order in space and first-order in time and a discussion of the accuracy along with a stability analysis can be found in Navon (1977).

(ii) *A.D.I. scheme.*

First we rewrite (42) as the first-order system

$$\frac{\partial \mathbf{w}}{\partial t} + A(\mathbf{w}) \frac{\partial \mathbf{w}}{\partial x} + B(\mathbf{w}) \frac{\partial \mathbf{w}}{\partial y} + C(\mathbf{w})\mathbf{w} = 0 \tag{46}$$

where

$$A = \begin{bmatrix} u & 0 & \Phi/2 \\ 0 & u & 0 \\ \Phi/2 & 0 & u \end{bmatrix}, \quad B = \begin{bmatrix} v & 0 & 0 \\ 0 & v & \Phi/2 \\ 0 & \Phi/2 & v \end{bmatrix}, \quad C = \begin{bmatrix} 0 & -f & 0 \\ f & 0 & 0 \\ 0 & 0 & 0 \end{bmatrix},$$

$\mathbf{w} = (u, v, \Phi)$. and $\Phi = 2\sqrt{(gh)}$. Now introduce the operators

$$P_{l,m}^n = -\frac{k}{2} [A(W_{l,m}^n) \frac{1}{2h} H_x + C_m^{(1)}]$$

$$Q_{l,m}^n = -\frac{k}{2} [B(W_{l,m}^n) \frac{1}{2h} H_y + C_m^{(2)}],$$

where

$$C_m^{(1)} = \begin{bmatrix} 0 & 0 & 0 \\ -f_m & 0 & 0 \\ 0 & 0 & 0 \end{bmatrix} \text{ and } C_m^{(2)} = \begin{bmatrix} 0 & f_m & 0 \\ 0 & 0 & 0 \\ 0 & 0 & 0 \end{bmatrix}.$$

The A.D.I. scheme introduced by Gustafsson (1971) is

$$(I - P_{l,m}^{n+1/2}) W_{l,m}^{n+1/2} = (1 + Q_{l,m}^n) W_{l,m}^n$$
$$(I - Q_{l,m}^{n+1}) W_{l,m}^{n+1} = (1 + P_{l,m}^{n+1/2}) W_{l,m}^{n+1/2}. \tag{47}$$

This scheme is shown to be stable and the solution of the non-linear system of algebraic equations at each time step by a quasi-Newton method is discussed by Gustafsson. A contrasting A.D.I. method based on a perturbation of a Crank–Nicolson type discretization is given by Fairweather and Navon (1979).

Exercise

4. Add the individual equations in (47) and show that the resulting difference formula represents (46) with second-order accuracy in time.

6.6 Free and moving boundary problems

One of the most important areas of application for finite difference methods is the numerical solution of *free-boundary* (F.B.) and *moving boundary* (M.B.) problems. The former usually occur in *stationary* problems modelled by an elliptic equation to be solved in a domain parts of whose boundary are unknown and are to be obtained as part of the solution. The latter occur in *time dependent* problems where the location of parts of the boundary evolves with the time dependent solution. The governing equation is usually parabolic but may be hyperbolic. Most F.B. and M.B. problems are extremely difficult to solve and *analytic* help should be used wherever possible. These problems occur in wide areas of practical interest including the metal, glass and plastics industries, oil exploration, diffusion and biological processes, meteorology etc. Useful surveys of the formulation and solution of free and moving boundary problems can be found in Cryer (1976) (F.B. problems only) and Furzeland (1977) (mainly M.B. problems). A recent alternative approach to the use of finite difference to solve these problems is that based on variational inequalities which are often solved numerically by finite element methods. Although this approach is particularly pleasing to mathematicians, it is still in an early stage of development, and no further mention of it will be made in this brief section. The interested reader is referred to Glowinski (1978) and Oden and Kikuchi (1979).

Rather than attempt a comprehensive treatment of problems and methods of solution we shall work through two problems (one F.B. and one M.B.) in some detail.

Seepage through a rectangular earth dam (F.B.)

The situation is illustrated in figure 4 where water seeps from an upstream reservoir (height H) through an earth dam into a lower reservoir (height L). The dam is sufficiently long to make the problem two-dimensional. Water *flows* only through the region R and the F.B. is the air-water interface denoted by Γ.

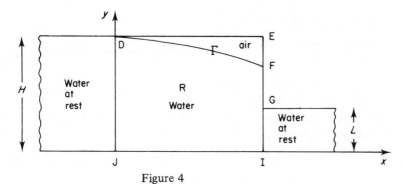

Figure 4

The mathematical model of this problem is:

Find u satisfying

$$\frac{\partial^2 u}{\partial x^2} + \frac{\partial^2 u}{\partial y^2} = 0 \quad \text{in } R \tag{48}$$

subject to the boundary conditions

$$u = -KH \qquad \text{on DJ}$$

$$\frac{\partial u}{\partial n} = 0 \qquad \text{on JI}$$

$$u = -KL \qquad \text{on IG} \tag{49}$$

$$u = -Ky \qquad \text{on GF}$$

$$\frac{\partial u}{\partial n} = 0 \qquad \text{on FD } (\Gamma),$$

and an *extra* boundary condition
$$u = -Ky \qquad \text{on FD } (\Gamma). \tag{50}$$

In the above the hydraulic head h is connected to the velocity potential u by Darcy's law

$$u = -Kh, \quad K \text{ constant.} \tag{51}$$

If the location of Γ were known, then (48) and (49) would suffice to determine u. Denoting the boundary of the closed region DJIGFD by ∂R, the problem given by (48), (49), and (50) can be reformulated as

$$Au = 0 \quad \text{in } R \tag{52}$$

subject to the boundary conditions

$$Bu = 0 \quad \text{on } \partial R,$$

$$Cu = 0 \quad \text{on } \Gamma, \tag{53}$$

where *the two boundary conditions on Γ can be interchanged*. This is the general structure of most F.B. problems.

We now turn to the solution of (52) and (53) and describe the *trial* F.B. method. This consists of two distinct steps.

(i) Obtain an approximation $U^{(k)}$ to the solution $u^{(k)}$ of the elliptic problem

$$Au^{(k)} = 0 \text{ in } R^{(k)}$$

$$Bu^{(k)} = 0 \text{ on } \partial R^{(k)} \tag{54}$$

where $k = 0, 1, 2, \ldots$.

(ii) From $\Gamma^{(k)}$ and $U^{(k)}$ compute a new F.B. $\Gamma^{(k+1)}$ by requiring that $CU^{(k)}$ should be (approximately) zero on $\Gamma^{(k+1)}$, i.e. move the boundary from $\Gamma^{(k)}$ to $\Gamma^{(k+1)}$.

Step (i) usually presents no great difficulty as the solution of (54) can be carried out using either finite difference (see Chapter 3) or finite element methods. Step (ii), however, is much more difficult and a successful strategy for this is essential. The main difficulty is that when $\Gamma^{(k+1)}$ is chosen to make $CU^{(k)}$ zero, the other boundary condition $BU^{(k)} = 0$ is no longer satisfied on $\Gamma^{(k+1)}$. The object is to construct boundary conditions $\hat{B}$ and $\hat{C}$ on Γ which are linear combinations of the original boundary conditions on Γ so that $\hat{B}u$ is less sensitive to movements of Γ. This was accomplished by Garabedian (1956) and a full description of his method can be found in Cryer (1976). In the present example, Garabedian's method demands that $\hat{B}$ be chosen so that

$$\frac{\partial}{\partial n} (\hat{B}u) = (\hat{B}u)_n = 0 \text{ on } \Gamma,$$

where $\partial/\partial n$ denotes differentiation along the normal to Γ. This gives

$$\hat{B}u = u_n - \frac{u_{nn}}{K\, y_n} (u + Ky) = 0$$

and

$$\hat{C}u = u + Ky = 0.$$

Since Γ and u are not known, the simplest strategy is to replace Γ by $\Gamma^{(k)}$ and so $U^{(k)}$ is defined to be the solution of the problem

$$AU^{(k)} = 0 \quad \text{in } R^{(k)}$$

$$\hat{B}^{(k)} U^{(k)} = U_n^{(k)} - \frac{U_{nn}^{(k)}}{K y_n} (U^{(k)} + Ky) = 0 \quad \text{on } \partial R^{(k)}$$

$$\hat{C}^{(k)} U^{(k)} = U^{(k)} + Ky = 0 \quad \text{on } \Gamma^{(k)}.$$

One-phase one-dimensional melting ice (M.B.)

The mathematical formulation of this problem is to find the unknowns, $u(x, t)$, the temperature distribution, and $s(t)$, the position of the ice—water interface, subject to the *parabolic* equation

$$\frac{\partial u}{\partial t} = \frac{\partial^2 u}{\partial x^2} \quad (0 < x < s(t), \quad t > 0), \tag{55}$$

the *initial* conditions

$$u = 0$$
$$\qquad\qquad \text{at } t = 0, \qquad\qquad\qquad\qquad (56)$$
$$s = 0$$

the *fixed* boundary condition

$$u = 1 \quad \text{on} \quad x = 0 \quad (t > 0), \qquad\qquad\qquad (57)$$

and the *moving* boundary conditions

$$u = 0$$
$$\qquad\qquad \text{on} \quad x = s(t) \quad (t > 0). \qquad\qquad (58)$$
$$\frac{\partial u}{\partial x} = -\frac{ds}{dt}.$$

This problem, known as the *Stefan* problem, is *one phase* if we assume that the semi-infinite block of ice $x > s(t)$ remains at the constant temperature $u = 0$.

There is no particular difficulty in writing down finite difference formulae to replace (55) and most of the *two-level* explicit and implicit schemes given in Chapter 2 will suffice. The problems all arise near the interface, the location of which is unknown for each advanced time level $t = (n + 1)k$, $n = 0, 1, 2, \ldots$. Its location is ùsually determined by some iterative process. Many devices have been tried to *front track* the moving boundary. These include varying the grid size in space or time, discretizing in time only to give two-point boundary problems at each time step, changing the roles of the dependent and independent variables, using the weak solution, and many others. Accounts of these with references can be found in Crank (1975) and Fox (1975). The implementation of these ideas to multi-phase problems in two and three space dimensions is complicated in the extreme with the moving boundary becoming more and more elusive. Conference proceedings which are relevant are edited by Ockendon and Hodgkins (1975) and Wilson et al. (1978).

6.7 Error growth in conduction—convection problems

In section 2.7, we discussed the matrix method of examining the stability of a difference calculation arising from a parabolic equation and stated that substantial error growth may occur with increasing n when the governing matrix is not symmetric even if the spectral radius condition $\rho(C) < 1$ is satisfied. The latter guarantees that the error will eventually tend to zero as $n \to \infty$.

In order to illustrate this point we consider the numerical solution of the model conduction—convection problem (cf. Chapter 2, case IV)

$$\frac{\partial u}{\partial t} = \frac{\partial^2 u}{\partial x^2} - b \frac{\partial u}{\partial x} \quad (b > 0) \qquad\qquad\qquad (59)$$

in the domain $0 < x < 1, t > 0$ with initial and boundary conditions

$$u(x, 0) = 0 \quad (0 < x < 1) \tag{60}$$

and

$$u(0, t) = 1, \quad \frac{\partial u}{\partial x}(1, t) = 0 \quad (t > 0), \tag{61}$$

respectively.

If we discretize in space, using central differences for the first derivative $\partial u/\partial x$ in (59) and (61), a system of ordinary differential equations is obtained, which after time discretization by forward differences gives

$$\mathbf{U}^{n+1} = C\mathbf{U}^n + r\mathbf{a} \quad (n = 0, 1, 2, \ldots) \tag{62}$$

where $\mathbf{U} = [U_1, U_2, \ldots, U_M]^T$, $\mathbf{a} = [1 + L, 0, \ldots, 0]^T$,

$$C = \begin{bmatrix} 1 - 2r & r(1 - L) & & & \\ r(1 + L) & 1 - 2r & r(1 - L) & & \\ & \cdot & \cdot & \cdot & \\ & & \cdot & \cdot & \cdot \\ & & r(1 + L) & 1 - 2r & r(1 - L) \\ & & & 2r & 1 - 2r \end{bmatrix}, \tag{63}$$

$r = k/h^2$, $L = \frac{1}{2}bh$, with grid spacings h and k in space and time respectively, and $Mh = 1$. The solution of (62) oscillates for $L > 1$ (Siemieniuch and Gladwell (1978)).

Using the notation and results of section 2.7, the error vector is $\mathbf{Z}^n (\equiv \mathbf{u}^n - \mathbf{U}^n)$ and the *necessary and sufficient* condition for $\|\mathbf{Z}^n\|$ to be bounded for all $n \geq 1$ is

$$\|C^n\| \leq K \tag{64}$$

where K is a constant independent of n and $\| \|$ is a suitable norm. We now examine the magnitude of the constant K when we use the *maximum norm* (see p. 11) to measure the error growth of difference calculations based on (62), which satisfy the spectral radius condition (2.43).

In order to set limits on the values of the pair (r, L) for our calculations, we note from (63) that

$$\|C\|_\infty = \begin{cases} |1 - 2r| + 2r & (0 \leq L \leq 1) \\ |1 - 2r| + 2rL & (L \geq 1) \end{cases}$$

and so $\|C\|_\infty \leq 1$ if and only if

$$0 < r \leq \frac{1}{2} \quad \text{and} \quad 0 \leq L \leq 1. \tag{65}$$

This contrasts sharply with the result

$$0 < r \leqslant \begin{cases} [1 + \sqrt{(1 - L^2)}]^{-1} & (0 \leqslant L \leqslant 1) \\ 1/L^2 & (L \geqslant 1) \end{cases} \tag{66}$$

obtained by Siemieniuch and Gladwell from the spectral radius condition (2.43). We are interested in the magnitude of $\|C^n\|_\infty$ for calculations based on values of (r,L) in the shaded region lying between the limits imposed by inequalities (65) and (66) (see figure 5). In all calculations, of course, the value of $\|C^n\|_\infty$ increases with n to a maximum value whereupon it decreases and eventually tends to zero as $n \to \infty$, consistent with the fact that $\rho(C) < 1$. A sample calculation showing the behaviour of $\|C^n\|_\infty$ with n and b for $r = 0.6$, $L = 1$ is shown in the following table.

b \ n	5	10	20	40	80	$\max_n \|C^n\|_\infty$
20	5.38	22.7	0.02	–	–	22.7
40	5.38	28.9	79.8	0.01	–	1030
80	5.38	28.9	837	698568	0.003	2.62×10^6

The entries in the final column show the maximum possible amplification of the initial error. A dash denotes entries which are negligibly small. Also as b increases (h decreases), the order of the matrix increases, in fact $M = \frac{1}{2}b$. Although this is a rather exceptional case of error growth, it illustrates the danger of accepting the spectral radius condition for explicit difference calculations of conduction–convection problems with significant first derivatives. A fuller study of this problem

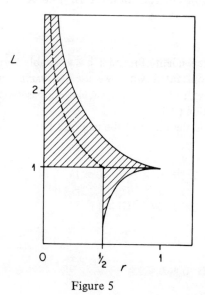

Figure 5

is contained in a paper by Griffiths, Christie, and Mitchell (1978) where an empirical stability limit

$$0 < r \leqslant \begin{cases} \tfrac{1}{2} & (0 \leqslant L \leqslant 1) \\[2mm] \tfrac{1}{2} L^{-2} & (L \geqslant 1) \end{cases} \tag{67}$$

is given. This is shown in figure 5 and lies 'between' the limits imposed by (65) and (66). A generalized upwind scheme for the first space derivative is also analysed in this paper.

Exercises

5. In the special case $L = 1, \tfrac{1}{2} \leqslant r \leqslant 1$, show that

$$\| C^n \|_{\infty} = (4r - 1)^n \qquad (n = 1, 2, \ldots, M - 1).$$

6. Ignoring boundary conditions, derive the difference approximation

$$U_m^{n+1} = (1 + r\delta_x^2 - rLH_x)U_m^n$$

for equation (59), and show that the von Neumann condition for stability is that given by (67). (This conforms with our statements in earlier chapters that the von Neumann method gives necessary conditions for convergence regardless of the type of boundary condition.)

References

Anderssen, R.S. and A.R.Mitchell. (1979). 'Analysis of generalised Galerkin methods in the numerical solution of elliptic equations', *Math. Meths. in the Appl. Sciences,* 1, 1-11.

Axelsson, O. (1972). 'A generalised SSOR method', *BIT* 13, 443-467.

Bramble, J.H. and B.E.Hubbard (1965). 'Approximation of solutions of mixed boundary value problems for Poisson's equation by finite differences', *J. Assoc. Comput. Mach.,* 12, 114-123.

Bramble, J.H., B.E.Hubbard and M.Zlamal (1968). 'Discrete analogues of the Dirichlet problem with isolated singularities', *SIAM J. Numer. Anal.,* 5, 1-25.

Broyden, C.G. (1975) .*Basic Matrices,* Macmillan, London.

Burns, A.M. (1978). 'A necessary condition for the stability of a difference approximation to a hyperbolic partial differential equation', *Maths. of Comp.,* 32, 707-724.

Cannon, J.R. and J. Douglas (1964). 'Three level alternating-direction iterative methods', Contrib. to *Diff. Equns.,* 3, 189-198.

Chorin, A.J. (1969). 'Numerical solution of the Navier-Stokes equations', *Maths. of Comp.,* 22, 745-762.

Christie, I. and A.R.Mitchell (1978). 'Upwinding of high order Galerkin methods in conduction-convection problems', *Int. J. Num. Meth. Eng.,* 12, 1764-1771.

Chu, C.K. and A. Sereny (1974). 'Boundary conditions in finite difference fluid dynamic codes', *J. Comp. Phys.,* 15, 476-491.

Ciment, M. and S.H.Leventhal (1975). 'High order compact implicit schemes for the wave equation', *Maths. of Comp.,* 29, 985-994.

Ciment, M., S.H.Leventhal and B.C.Weinberg (1978). 'The operator compact implicit method for parabolic equations', *J. Comp. Phys.,* 28, 135-166.

Cole, J.D. (1951). 'On a quasi-linear parabolic equation occuring in aerodynamics', *Quart. of App. Maths.,* 9, 225-236.

Collatz, L. (1960). *Numerical Treatment of Differential Equations,* Springer, Berlin (3rd edition).

Courant, R., K.Friedrichs and H.Lewy (1928). 'Über die partiellen Differenzengleichungen der mathematischen Physik', *Mathematische Annalen,* 100, 32-74.

Courant, R. and D.Hilbert (1962). *Methods of Mathematical Physics,* Interscience, New York.

Crank, J. (1975). 'Finite difference methods' in *Moving Boundary Problems in Heat Flow and Diffusion* (ed. J.E.Ockendon and W.R.Hodgkins), Clarendon Press, Oxford, 192-207.

Crank, J.and R.M.Furzeland(1977). 'The numerical solution of elliptic and parabolic partial differential equations with boundary singularities', Report No.TR/68

(Revised), Brunel University.

Cryer, C.W. (1976). 'A survey of trial free-boundary methods for the numerical solution of free-boundary problems', M.R.C. Report 1693, University of Wisconsin.

Cullen, M.J.P. (1974). 'A finite element method for a non-linear initial value problem', *J. Inst. Maths. Applics.*, **13**, 233-247.

Douglas, J. and H.H.Rachford (1956). 'On the numerical solution of heat conduction problems in two and three space variables', *Trans. Amer. Math. Soc.*, **82**, 421-439.

Douglas, J. (1962). Alternating direction methods for three space variables', *Numer. Math.*, **4**, 41-63.

Douglas, J. and B.F.Jones (1963). 'On predictor-corrector methods for non-linear parabolic differential equations', *J. Soc. Ind. Appl. Math.*, **11**, 195-204.

Du Fort, E.C. and S.P.Frankel (1953). 'Stability conditions in the numerical treatment of parabolic differential equations', *M.T.A.C.*, **7**, 135.

Dupont, T., R.P.Kendall and H.H.Rachford (1968). 'An approximate factorization procedure for solving self-adjoint elliptic difference equations', *SIAM J. Numer. Anal.*, **5**, 559-573.

D'Yakonov, Ye G. (1963). 'On the application of disintegrating difference operators', *Z, Vycisl. Mat. i. Mat. Fiz.*,**3**, 385-388.

Fadeeva, V.N. (1959). *Computational Methods of Linear Algebra*, Dover Press, New York.

Fairweather, G. and A.R.Mitchell (1965). 'A high accuracy alternating direction method for the wave equation', *J. Inst. Math. Applics.*, **1**, 309-316.

Fairweather, G. and A.R.Mitchell (1967). 'A new computational procedure for A.D.I. methods', *SIAM J. Numer. Anal.*, **4**, 163-170.

Fairweather, G. (1978). *Finite Element Galerkin Methods for Differential Equations*, Marcel Dekker, Basel.

Fairweather, G. and I.M.Navon (1979). 'A linear A.D.I. method for solving the shallow-water equations', *J. Comp. Phys.* (to appear).

Fornberg, B. (1973). 'On the instability of Leap-Frog and Crank-Nicolson approximations of a non-linear partial differential equation', *Math. Comp.*, **27**, 45-57.

Fox, L. (1964). *Introduction to Numerical Linear Algebra*, Oxford University Press, Oxford.

Fox, L., P.Henrici and C.Moler (1967). 'Approximations and bounds for eigenvalues of elliptic operators', *SIAM J. Numer. Anal.*, **4**, 89-102.

Fox, L. and R.Sankar (1969). 'Boudary singularities in linear elliptic differential equations', *J. Inst. Maths. Applics.*, **5**, 340-350.

Fox, L. (1975). 'What are the best numerical methods?' in *Moving Boundary Problems in Heat Flow and Diffusion* (ed. J.R.Ockendon and W.R.Hodgkins) Clarendon Press, Oxford 210-241.

Furzeland, R.M. (1977). 'A survey of the formulation and solution of free and moving boundary (Stefan) problems', Brunel University TR/76.

Gane, C.R. and A.R.Gourlay (1977). 'Block hopscotch procedures for second order parabolic differential equations', *J. Inst. Maths. Applics.*, **19**, 205-216.

Garabedian, P.R. (1956). 'The mathematical theory of three dimensional cavaties and jets', *Bull. Amer. Math. Soc.*, **62**, 219-235.

Garabedian, P.R. (1964). *Partial Differential Equations*, John Wiley, New York.

Gear, C.W. (1971). *Numerical Initial Value Problems in Ordinary Differential Equations*, Prentice Hall, Englewood Cliffs, N.J.

George, A., W.G.Poole Jnr and R.G.Voigt (1978). 'Incomplete nested dissection for solving n by n grid problems', *SIAM J. Numer. Anal.*, **15**, 662-673.

Glowinski, R. (1978). 'Finite elements and variational inequalities, IRIA Laboratory Report 78010.

Godunov, S.K. and V.S.Ryabenkii (1963). 'Spectral stability criteria of boundary value problems for non-selfadjoint difference equations', *Rus. Math. Surv.*, **18**, 1-12.

Gordon, P. (1965). 'Nonsymmetric difference equations', *J. Soc. Ind. Appl. Math.*, **13**, 667-673.

Gottlieb, D. and E.Turkel (1974). 'Phase error and stability of second order methods for hyperbolic problems II, *J. Comp. Phys.*, **15**, 251-265.

Gourlay, A.R. and J.Ll.Morris (1968). 'A multistep formulation of the optimised Lax Wendroff method for non-linear hyperbolic systems in two space variables', *Maths. of Comp.*, **22**, 715-720.

Gourlay, A.R. and A.R.Mitchell (1969). 'The equivalence of certain alternating direction and locally one-dimensional difference methods', *SIAM J. Numer. Anal.*, **6**, 37-46.

Gourlay, A.R. (1970). 'Hopscotch: A fast second-order partial differential equation solver', *J. Inst. Math. App.*, **6**, 375-390.

Gourlay, A.R. and G.R.McGuire (1971). 'General hopscotch algorithm for the numerical solution of partial differential equations', *J. Inst. Maths. Applics.*, **7**, 216-227.

Gourlay, A.R. and G.A.Watson (1973). *Computational Methods for Matrix Eigenproblems*, John Wiley, New York.

Gourlay, A.R. and S.McKee (1977). 'The construction of hopscotch methods for parabolic and elliptic equations in two space dimensions with a mixed derivative', *J. Comp. Appl. Maths.*, **3**, 201-206.

Gourlay, A.R. (1977). 'Splitting methods for time dependent partial differential equations' in *The State of the Art in Numerical Analysis*, (ed. D.A.H.Jacobs) Academic Press, London.

Greenspan, D. and P.C.Jain (1964). 'On non-negative difference analogues of elliptic differential equations', Technical Report No.490, Mathematics Research Center, Madison.

Greig, I.S. and J.Ll.Morris (1976). 'A hopscotch method for the Kortweg-de-Vries equation', *J. Comp. Phys.*, **20**, 64-80.

Gresho, P.M., R.L. Lee and R.L. Sani (1978), 'Advection-dominated flows with emphasis on the consequences of mass-lumping', in *Finite Elements in Fluids* Vol. 3, 335-350, John Wiley and Sons, New York.

Griffiths, D.F. (1977). 'A numerical study of a singular elliptic boundary value problem', *J. Inst. Maths. Applics.*, **19**, 59-69.

Griffiths, D.F., I.Christie and A.R.Mitchell (1978). 'Analysis of error growth for explicit difference schemes in conduction-convection problems', University of Dundee Report NA/29.

Gustafsson, B. (1971). 'An alternating direction implicit method for solving the shallow-water equations', *J. Comp. Phys.*, **7**, 239-254.

Gustafsson, B., H-O.Kreiss and A.Sundström (1972). 'Stability theory of difference approximations for mixed initial boundary value problems II', *Maths. of Comp.*, **26**, 649-686.

Heinrich, J.C., P.S.Huyakorn, A.R.Mitchell and O.C.Zienkiewicz (1977). 'An upwind finite element scheme for two dimensional convective transport equations', *Int. J. Num. Meth. Eng.*, **11**, 131-144.

Hestenes, M.R. and E.Stiefel (1952). 'Methods of conjugate gradients for solving linear systems', *J. Res. Nat. Bur. Standards*, **49**, 33-53.

Hirsh, R.S. (1975). 'Higher order accurate difference solutions of fluid mechanics problems by a compact differencing technique', *J. Comp. Phys.*, **19**, 90-109.

Hubbard, B. (1966). 'Some locally one-dimensional difference schemes for parabolic equations in an arbitrary region', *Maths. of Comp.*, **20**, 53-59.

Il'in, A.M. (1969). 'Differencing scheme for a differential equation with a small parameter affecting the highest derivative', *Math. Notes Acad. Sc. USSR*, **6**, 596-602.

Isaacson, E. and H.B.Keller (1967). *Analysis of Numerical Methods*, John Wiley, New York.

Jacobs, D.A.H. (1971). 'The hopscotch method applied to the Navier Stokes equations', Central Electricity Research Laboratories RD/L/N 164/71.

Jeffrey, A. and T.Tanuiti (1964). *Non-Linear Wave Propagation*, Academic Press, New York.

Kantarovich, L. and V.Krylov (1958). *Approximate Methods of Higher Analysis*, Noordhoff, Gronigen.

Keast, P. and A.R.Mitchell (1967). 'Finite difference solution of the third boundary problem in elliptic and parabolic equations', *Numer. Math.*, **10**, 67-75.

Kershaw, D.S. (1978). 'The incomplete Cholesky-conjugate gradient method for the iterative solution of systems of linear equations', *J. Comp. Phys.*, **26**, 43-65.

Kreiss, H-O. and E.Lundqvist (1968). 'On difference approximations with wrong boundary values', *Maths. of Comp.*, **22**, 1-12.

Kreiss, H-O. (1968). 'Stability theory for difference approximation of mixed initial boundary value problems I', *Maths. of Comp.*, **22**, 703-714.

Kreiss, H-O. and J.Oliger (1972). 'Comparison of accurate methods for the integration of hyperbolic equations', *Tellus*, 24, 199-215.

Lambert, J.D. (1972). *Computational Methods in Ordinary Differential Equations*, John Wiley, New York.

Lees, M. (1962). 'Alternating direction methods for hyperbolic differential equations', *J. Soc. Indust. App. Math.*, **10**, 610-616.

Lees, M. (1966). 'A linear three level difference scheme for quasilinear parabolic equations', *Maths. of Comp.*, **20**, 516-522.

MacCormack, R.W. (1969). 'The effect of viscosity in hypervelocity impact cratering', AIAA Paper No.69-354.

MacCormack, R.W. (1971). 'Numerical solution of the interaction of a shock wave with a laminar boundary layer', *Proceedings of the Second International Conference on Numerical Methods in Fluid Dynamics*, (ed. M.Holt) Lecture Notes in Physics, Vol. 8, Springer-Verlag.

McGuire, G.R. and J.Ll.Morris (1973). 'A class of second-order accurate methods for the solution of systems of conservation laws', *J. Comp. Phys.*, **11**, 531-539.

McKee, S. and A.R.Mitchell (1970). 'Alternating direction methods for parabolic equations in two space dimensions with a mixed derivative', *Computer Journal*, **13**, 81-86.

McKee, S. (1973). 'High accuracy A.D.I. methods for hyperbolic equations with variable coefficients', *J. Inst. Maths. Applics.*, **11**, 105-109.

McLaurin, J.W. (1974). 'A general coupled equation approach for solving the biharmonic boundary value problem', *SIAM J. Numer. Anal.*, **11**, 14-33.

May, T. (1978). 'Boundary conditions in the numerical intergration of hyperbolic equations', Ph.D. Thesis, University of Reading.

Meijerink, J.A. and H.A. van der Vorst (1977). 'An iterative solution method for linear systems of which the coefficient matrix is a symmetric M-matrix', *Maths. Comp.*, **31**, 148-162.

Mitchell, A.R. and R.P.Pearce (1963). 'Explicit difference methods for solving the cylindrical heat conduction equation', *Maths. of Comp.*, **17**, 426-432.

Mitchell, A.R. and G.Fairweather (1964). 'Improved forms of the alternating direction methods of Douglas, Peaceman and Rachford for solving parabolic and elliptic equations', *Numer. Maths.*, **6**, 285-292.

Mitchell, A.R. (1971). 'Splitting methods in partial differential equations', *Abhand-*

lungen aus dem Mathematischen Seminar der Universität Hamburg, **36**, 45-56.

Mitchell, A.R. and R.Wait (1977). *The Finite Element Method in Partial Differential Equations,* John Wiley, New York.

Morris, J.Ll. (1978). Private communication.

Morton, K.W. (1977). 'Initial-value problems by finite difference and other methods', in *The State of the Art in Numerical Analysis* (ed. D.A.H.Jacobs) Academic Press, London.

Morton, K.W. (1979). 'Stability of finite difference approximations to a diffusion-convection equation', University of Reading, Report 3/79.

Navon, I.M. (1977). 'Application of a new partly-implicit time differencing scheme for solving the shallow-water equations', C.S.I.R. report WISK 278, Pretoria.

Ockendon, J.R. and W.R.Hodgkins (1975). *Moving Boundary Problems in Heat Flow and Diffusion,* Clarendon Press, Oxford.

Oden, J.T. and N. Kikuchi (1979). 'Theory of variational inequalities', TICOM Report 79-4, University of Texas at Austin.

Ortega, J.M. and W.C.Rheinboldt (1970). *Iterative Solutions of Nonlinear Equations in Several Variables,* Academic Press, New York.

Peaceman, D.W. and H.H.Rachford (1955). 'The numerical solution of parabolic and elliptic differential equations', *J. Soc. Indust. App. Math.,* **3**, 28-41.

Pearson, C.E. (1965). 'A computational procedure for viscous flow problems', *J. Fluid Mech.,* **21**, 611-622.

Peyret, R. and H.Viviand (1975). 'Computation of viscous compressible flows based on the Navier-Stokes equations'. Advisory Group for Aerospace Research and Development report AGARD-AG-212.

Polak, S.J. (1974). 'An increased accuracy scheme for diffusion equations in cylindrical co-ordinates', *J. Inst. Maths. Applics.,* **14**, 197-201.

Reid, J.K. (1972). 'The use of conjugate gradients for systems of linear equations possessing "Property A" '. *SIAM J. Numer. Anal.,* **9**, 325-332.

Richtmyer, R.D. (1963). 'A survey of difference methods for non-steady fluid dynamics', NCAR Technical Note 63-2, Boulder, Colorado.

Richtmyer, R.D. and K.W.Morton (1967). *Difference Methods for Initial Value Problems,* John Wiley, New York.

Roache,P.J. (1972). *Computational Fluid Dynamics,* Hermosa Publishers, Albuquerque, New Mexico.

Rubin, E.L. and S.Z.Burstein (1967). 'Difference methods for the inviscid and viscous equations of a compressible gas', *J. Comp. Phys.,* **2**, 178-196.

Samarskii, A.A. (1964a). 'Local one dimensional difference schemes for multi-dimensional hyperbolic equations in an abitrary region', *Z. Vycisl. Mat. i Mat. Fiz.,* **4**, 21-35.

Samarskii, A.A. (1964b). 'An accurate high order difference scheme for a heat conductivity equation with several space variables', *Z. Vycisl. Mat. i Mat. Fiz.,* **4**, 222-223.

Samarskii, A.A. (1964c). 'Economical difference schemes for parabolic equations with mixed derivatives', *Z. Vycisl. Mat. i Mat. Fiz.,* **4**, 753-759.

Siemieniuch, J.L. and I.Gladwell (1978). 'Analysis of explicit difference methods for a diffusion-convection equation', *Int. J. Num. Meth. Engng.,* **12**, 899-916.

Smith, J. (1968). 'The coupled equation approach to the numerical solution of the biharmonic equation by finite differences. I', *SIAM J. Numer. Anal.,* **5**, 323-339.

Stewart, G.W. (1973). *Introduction to Matrix Computations,* Academic Press, New York.

Stone, H.L. (1968). 'Iterative solution of implicit approximations of multi-dimensional partial differential equations', *SIAM J. Numer. Anal.,* **5**, 530-558.

Strang, W.G. (1963). 'Accurate partial difference methods I: Linear Cauchy problems', *Arch. Rat. Mech. Anal.,* **12**, 392-402.

Strang, W.G. (1964). 'Accurate partial difference methods II: Non-linear problems', *Numer.Math.*, **6**, 37-46.

Strang, W.G. (1968). 'On the construction and comparison of difference schemes', *SIAM J. Numer. Anal.*, **5**, 506-517.

Sundström, A. (1975). 'Note on the paper: "Boundary conditions in finite difference fluid dynamics codes" by C.K.Chu and A.Sereny', *J. Comp. Phys.*, **17**, 450-454.

Tee, G.J. (1963). 'A novel finite-difference approximation to the biharmonic operator', *Computer Journal*, **6**, 177-192.

Tikhonov,A.N. and A.A.Samarskii (1961). 'Homogeneous difference schemes', *Z. Vycisl. Mat. i Mat. Fiz.*, **1**, 5-63.

Todd, J. (1962). *Survey of Numerical Methods*, McGraw-Hill, New York.

Turkel, E. (1974). 'Phase error and stability of second order methods for hyperbolic problems I', *J. Comp. Phys.*, **15**, 226-250.

Varga, R.S. (1962). *Matrix Iterative Analysis*, Prentice Hall, Englewood Cliffs, N.J.

Wachspress, E.L. (1957). 'A generalized two space dimension multigroup coding for the IBM 704', CURE report KAPL-1724, General Electric Co., New York.

Wachspress, E.L. (1966). *Iterative Solution of Elliptic Systems*, Prentice-Hall, Englewood Cliffs, N.J.

Wachspress, E.L. (1977). 'Isojacobic crosswind differencing'. *Lecture Notes in Mathematics,* **630** (ed. G.A.Watson) 190-199, Springer-Verlag, Berlin.

Wendroff, B. (1960). 'On centred difference equations for hyperbolic systems', *J. Soc. Indust. App. Math.*, **8**, 549-555.

Widlund, O.B. (1966). 'On the rate of convergence of an alternating direction implicit method in the non-commutative case', *Maths. of Comp.*, **20**, 500-515.

Wilkinson, J.H. (1965). *The Algebraic Eigenvalue Problem*, Oxford University Press, Oxford.

Wilson, D.G., A.D. Solomon and P.T. Boggs (1978). *Moving Boundary Problems,* Academic Press, New York.

Wilson,J.C. (1972). 'Stability of Richtmyer type difference schemes in any finite number of space variables and their comparison with multistep Strang schemes', *J. Inst. Maths. Applics.*, **10**, 238-257.

Wood, W.L. and R.W.Lewis (1975). 'A comparison of time marching schemes for the transient heat conduction equation', *Int. J. Num. Meth. Engng.*, **9**, 679-689.

Young, D.M. (1971). *Iterative Solution of Large Linear Systems*, Academic Press, New York.

Zlamal, M. (1967). 'Discretization and error estimates for elliptic boundary value problems of the fourth order', *SIAM J. Numer. Anal.*, **4**, 626-639.

Index